MATHEMATICAL CONCEPTS FOR MECHANICAL ENGINEERING DESIGN

MATHEMATICAL CONCEPTS FOR MECHANICAL ENGINEERING DESIGN

Kaveh Hariri Asli, PhD, Hossein Sahleh, PhD, and
Soltan Ali Ogli Aliyev, PhD

Apple Academic Press

TORONTO   NEW JERSEY

Apple Academic Press Inc. | Apple Academic Press Inc.
3333 Mistwell Crescent | 9 Spinnaker Way
Oakville, ON L6L 0A2 | Waretown, NJ 08758
Canada | USA

©2014 by Apple Academic Press, Inc.

First issued in paperback 2021

Exclusive worldwide distribution by CRC Press, a member of Taylor & Francis Group

No claim to original U.S. Government works

ISBN 13: 978-1-77463-291-8 (pbk)
ISBN 13: 978-1-926895-62-8 (hbk)

Library of Congress Control Number: 2013951201

Library and Archives Canada Cataloguing in Publication

Asli, Kaveh Hariri, author
Mathematical concepts for mechanical engineering design/Kaveh Hariri Asli, PhD, Hossein Sahleh, PhD, and Soltan Ali Ogli Aliyev, PhD.

Includes bibliographical references and index.
ISBN 978-1-926895-62-8
1. Fluid mechanics--Mathematical models. 2. Mechanical engineering--Mathematical models. 3. Mechanical engineering--Data processing.
I. Sahleh, Hossein, 1952-, author II. Aliyev, Soltan Ali Ogli, author III. Title.

QA901.A86 2013   532.01'5118   C2013-906695-0

Apple Academic Press also publishes its books in a variety of electronic formats. Some content that appears in print may not be available in electronic format. For information about Apple Academic Press products, visit our website at **www.appleacademicpress.com** and the CRC Press website at **www.crcpress.com**

ABOUT THE AUTHORS

Kaveh Hariri Asli, PhD

Kaveh Hariri Asli, PhD, is a professional mechanical engineer with over 30 years of experience in practicing mechanical engineering design and teaching. He is the author of over 50 articles and reports in the fields of fluid mechanics, hydraulics, automation, and control systems. Dr. Hariri has consulted for a number of major corporations.

Hossein Sahleh, PhD

Hossein Sahleh, PhD, is a university lecturer with 30 years of experience in teaching and research in mathematics. He is the author of many papers in journals and conference proceedings and is an editorial board member of several journals.

Soltan Ali Ogli Aliyev, PhD

Soltan Ali Ogli Aliyev, PhD, is Deputy Director of the Department of Mathematics and Mechanics at the National Academy of Science of Azerbaijan (AMEA) in Baku, Azerbaijan. He served as a professor at several universities. He is the author and editor of several book as well as of a number of papers published in various journals and conference proceedings.

ABOUT THE AUTHORS

CONTENTS

LIST OF ABBREVIATIONS

FD	Finite differences
FE	Finite elements
FV	Finite volume
FVM	Finite volume method
MOC	Method of characteristics
PLC	Program logic control
RTC	Real-time control
WCM	Wave characteristic method

LIST OF SYMBOLS

V = water flow or discharge $\left(m^3/s\right), \left(lit/s\right)$

C = the wave velocity $\left(m/s\right)$

E_{∞} = modulus of elasticity of the liquid (water), $MR = a + bt + ct^2$, $\left(kg/m^2\right)$

E = modulus of elasticity for pipeline material Steel, $D_{EFF} = \alpha_F D_{AV}$

d = outer diameter of the pipe (m)

δ = wall thickness (m)

V_0 = liquid with an average speed $\left(m/s\right)$

T = time (s)

h_0 = ordinate denotes the free surface of the liquid (m)

u = fluid velocity $\left(m/s\right)$

λ = wavelength

$(hu)_x$ = amplitude a

$\dfrac{\partial h}{\partial t} dx$ = changingthe volume of fluid between planes in a unit time

h_0 = phase velocity $\left(m/s\right)$

V_{Φ} = expressed in terms of frequency

f = angular frequency

ω = wave number

Φ = a function of frequency and wave vector

$v_{\Phi}(k)$ = phase velocity or the velocity of phase fluctuations $\left(m/s\right)$

$\lambda(k)$ = wavelength

k = waves with a uniform length, but a time-varying amplitude

$k_{**}(\omega)$ = damping vibrations in length

ω = waves with stationary in time but varying in length amplitudes

P_{si0}= saturated vapor pressure of the components of the mixture at an initial temperature of the mixture T_0,(pa)

μ_2, μ_1 = molecular weight of the liquid components of the mixture

B = universal gas constant

P_i = the vapor pressure inside the bubble (pa)

T_{ki} =temperature evaporating the liquid components (°C)

l_i= specific heat of vaporization

D = diffusion coefficient volatility of the components

N_{k_0}, N_{c_0} = molar concentration of 1-th component in the liquid and steam

c_l =the specific heats of liquid

a_l = vapor at constant pressure

a_l = thermal diffusivity

ρ_v = vapor density $\left(\dfrac{kg}{m^3}\right)$

$R = r = R(t)$= radius of the bubble (m)

λ_l = coefficient of thermal conductivity

ΔT = overheating of the liquid (°C)

β = is positive and has a pronounced maximum at $k_0 = 0,02$

P_1 and P_2= the pressure component vapor in the bubble (pa)

P_∞ = the pressure of the liquid away from the bubble (pa)

σ = surface tension coefficient of the liquid

V_1= kinematic viscosity of the liquid

k_R = the concentration of the first component at the interface

n_i= the number of moles

V = volume $\left(m^3\right)$

B = gas constant

T_v = the temperature of steam $(°C)$

P_i' = the density of the mixture components in the vapor bubble $\left(kg/m^3\right)$

μ_i = molecular weight

p_{si} = saturation pressure (pa)

l_i = specific heat of vaporization

k = the concentration of dissolved gas in liquid

v_ϕ = speed of long waves

h = liquid level is above the bottom of the channel

ξ = difference of free surface of the liquid and the liquid level is above the bottom of the channel (a deviation from the level of the liquid free surface)

u = fluid velocity $\left(m/s\right)$

τ = time period

a = distance of the order of the amplitude

k = wave number

$v_\phi(k)$ = phase velocity or the velocity of phase fluctuations

$\lambda(k)$ = wave length

$\omega_{**}(k)$ = damping the oscillations in time

λ = coefficient of combination

q = flow rate $\left(m^3/s\right)$

μ = fluid dynamic viscosity $\left(kg/m.s\right)$

γ = specific weight $\left(N/m^3\right)$

j = junction point (m)

y = surgetank and reservoir elevation difference (m)

k = volumetriccoefficient $\left(GN/m^2\right)$

T = period of motion

A = pipe cross-sectional area $\left(m^2\right)$

dp = static pressure rise (m)

h_p = head gain from a pump (m)

h_L = combined head loss (m)

E_v = bulk modulus of elasticity (pa), $\left(\frac{kg}{m^2}\right)$

α = kinetic energy correction factor

P = surge pressure (pa)

g = acceleration of gravity $\left(\frac{m}{s^2}\right)$

K = wave number

T_p = pipe thickness (m)

E_p = pipe module of elasticity, (pa) $\left(\frac{kg}{m^2}\right)$

E_W = module of elasticity of water (pa), $\left(\frac{kg}{m^2}\right)$

C_1 = pipe support coefficient

$Y\max = Max.$ Fluctuation

R_0 = radiuses of a bubble (m)

D = diffusion factor

β = cardinal influence of componential structure of a mixture

N_{k_0}, N_{c_0} = mole concentration of 1-th component in a liquid and steam

γ = Adiabatic curve indicator

c_p c_{pv} = specific thermal capacities of a liquid at constant pressure

a_l = thermal conductivity factor

ρ_v = steam density $\left(\frac{kg}{m^3}\right)$

R = vial radius (m)

λ_l = heat conductivity factor

k_0 = values of concentration, therefore

w_l = velocity of a liquid on a bubble surface $\left(m\middle/s\right)$

p_1 and p_2 = pressure steam component in a bubble (pa)

P_∞ = pressure of a liquid far from a bubble (pa)

σ and v_1 = factor of a superficial tension of kinematics viscosity of a liquid

B = gas constant

T_v = temperature of a mixture (°C)

ρ_i' = density a component of a mix of steam in a bubble $\left({}^{kg}\!/_{m^3} \right)$

μ_i = molecular weight

j_i = the stream weight

i = components from an ($i = 1,2$) inter-phase surface in $r = R(t)$

W_i = diffusion speeds of a component on a bubble surface $\left({}^{m}\!/_{s} \right)$

l_i = specific warmth of steam formation

k_R = concentration 1-th components on an interface of phases

T_0, T_{ki} = liquid components boiling temperatures of a binary mixture at initial pressure p_0, (°C)

D = diffusion factor

λ_l = heat conductivity factor

Nu_l= parameter of Nusselt

a_l = thermal conductivity of liquids

c_l = factor of a specific thermal capacity

P_{el} = Number of Pekle

Sh = parameter of Shervud

pe_d= diffusion number the Pekle

ρ = density of the binary mix $\left({}^{kg}\!/_{m^3} \right)$

t = time (s)

λ_0 = unitof length

V = velocity $\left(m \,/_s \right)$

S = length (m)

D= diameterof each pipe (m)

R = piperadius (m)

ν = fluiddynamic viscosity $\left(\dfrac{kg}{m.s}\right)$

h_p = head gain from a pump (m)

h_L = combinedhead loss (m)

C = velocityof surge wave $\left(\dfrac{m}{s}\right)$

$P/_\gamma$ = pressurehead (m)

Z = elevationhead (m)

$V^2/_{2g}$ = velocityhead (m)

γ = specific weight $\left(\dfrac{N}{m^3}\right)$

Z = elevation (m)

H_P = surgewave head at intersection points of characteristic lines (m)

V_P = surgewave velocity at pipeline points- intersection points of characteristic lines $\left(\dfrac{m}{s}\right)$

V_{ri} = surgewave velocity at right hand side of intersection points of characteristic lines $\left(\dfrac{m}{s}\right)$

H_{ri} = surgewave head at right hand side of intersection points of characteristic lines (m)

V_{le} = surgewave velocity at left hand side of intersection points of characteristic lines $\left(\dfrac{m}{s}\right)$

H_{le} = surgewave head at left hand side of intersection points of characteristic lines (m)

p = pressure (bar), $\left(\dfrac{N}{m^2}\right)$

dv = incrementalchange in liquid volume with respect to initial volume

$\left(\dfrac{d\rho}{\rho}\right)$ = incremental change in liquid density with respect to initial density

SUPERSCRIPTS

C^- = characteristic lines with negative slope

C^+ = characteristic lines with positive slope

SUBSCRIPTS

Min. = Minimum

Max. = Maximum

Lab. = Laboratory

MOC = Method of Characteristic

PLC = Program Logic Control

PREFACE

In this book a computational and practical method was used for a prediction of mechanical systems failure. The proposed method allowed for any arbitrary combination of devices in mechanics of a liquid, gas and plasma. A scale model and a prototype (real) system were used. This book presents the performances of a computational method for system failure prediction by numerical analysis and nonlinear dynamic model. In this book various methods were developed to solve fluid mechanics problems. This range includes the approximate equations to numerical solutions of the nonlinear Navier–Stokes equations. The model was presented by method of the Eulerian based expressed in a method of characteristics (MOC): finite deference, finite volume, and finite element. It was defined by finite difference form for heterogeneous model with varying state in the system. This book offers MOC as a computational approach from theory to practice in numerical analysis modeling. Therefore, it was presented as the mathematical concepts for mechanical engineering design and as a computationally efficient method for flow irreversibility prediction in a practical case.

This book includes the research of the authors on the development of optimal mathematical models. The problem was presented by means of theoretical and experimental research. The authors also used modern computer technology and mathematical methods for analysis of nonlinear dynamic processes. This collection develops a new method for the calculation of mathematical models by computer technology. The process of entering input for the calculation of mathematical models was simplified for the user through the use of advances in control and automation of mechanical systems. The authors used parametric modeling technique and multiple analyzes for mechanical systems. This method has provided a suitable way for detecting, analyzing, and recording mechanical systems fault. Certainly, it can be assumed as a method with high-speed response ability for detecting the failure phenomena during irregular condition. The authors believe that the results of this book have a new idea and it can help to reduce the risk of system damage or failure at the mechanical systems.

— **Kaveh Hariri Asli, PhD, Hossein Sahleh, PhD, and Soltan Ali Ogli Aliyev, PhD**

INTRODUCTION

This book uses many computational methods for mechanical engineering design. Proposed methods allowed for any arbitrary combination of devices in system. Methods are used by scale models and prototype system.

In this book a computational and practical method was used for prediction of system failure. The proposed method allowed for any arbitrary combination of devices in fluid mechanics system and heat and mass transfer rates. A scale model and a prototype (real) system were used for mechanics of a liquid, gas and plasma. This book presents the performances of computational method for prediction of water distribution failure by application of numerical analysis and nonlinear dynamic modeling. In this book various methods were developed to solve water flow failure in mechanical systems.

This book includes the research of the authors on the development of optimal mathematical models. In order to predict urban water system failure, the propagation of the fluid movements in the pipeline and conducting numerical experiments to assess the adequacy of the proposed model were performed. The problem was presented by means of theoretical and experimental research. The authors also used modern computer technology and mathematical methods for analysis of nonlinear dynamic processes. This collection develops a new method for the calculation and prediction, for example, about high air velocities that will reduce the thickness of the stationary gas film on the surface of the solids and hence increase the heat and mass transfer coefficients. In practical designing it is found to be more reliable to consider heat transfer rates than mass transfer rates, as the latter are a function of surface temperature of the wet solid that is difficult to determine and cannot, in practice, be assumed to be that of the wet-bulb temperature of the air with an adequate degree of accuracy.

CHAPTER 1

HEAT FLOW—FROM THEORY TO PRACTICE

CONTENTS

1.1 INTRODUCTION

When faced with a drying problem on an industrial scale, many factors have to be taken into account in selecting the most suitable type of dryer to install and the problem requires to be analyzed from several standpoints. Even an initial analysis of the possibilities must be backed up by pilot-scale tests unless previous experience has indicated the type most likely to be suitable. The accent today, due to high labor costs, is on continuously operating unit equipment, to what extent possible automatically controlled. In any event, the selection of a suitable dryer should be made in two stages, a preliminary selection based on the general nature of the problem and the textile material to be handled, followed by a final selection based on pilot-scale tests or previous experience combined with economic considerations [1-5].

A leather industry involves a crucial energy-intensive drying stage at the end of the process to remove moisture left from dye setting. Determining drying characteristics for leather, such as temperature levels, transition times, total drying times, and evaporation rates, is vitally important so as to optimize the drying stage. Meanwhile, a textile material undergoes some physical and chemical changes that can affect the final leather quality [6-11].

In considering a drying problem, it is important to establish at the earliest stage, the final or residual moisture content of the textile material, which can be accepted. This is important in many hygroscopic materials and if dried below certain moisture content they will absorb or "regain" moisture from the surrounding atmosphere depending upon its moisture and humidity. The material will establish a condition in equilibrium with this atmosphere and the moisture content of the material under this condition is termed the equilibrium moisture content. Equilibrium moisture content is not greatly affected at the lower end of the atmospheric scale but as this temperature increases the equilibrium moisture content figure decreases, which explains why materials can in fact be dried in the presence of superheated moisture vapor. Meanwhile, drying medium temperatures and humidities assume considerable importance in the operation of direct dryers [12-21].

It should be noted that two processes occur simultaneously during the thermal process of drying a wet leather material, namely, heat transfer in order to raise temperature of the wet leather and to evaporate its moisture content together with mass transfer of moisture to the surface of the textile material and its evaporation from the surface to the surrounding atmosphere which, in convection dryers, is the drying medium. The quantity of air required to remove the moisture as liberated, as distinct from the quantity of air which will release the required amount of heat through a drop in its temperature in the course of drying, however, has to be determined from the known capacity of air to pick up moisture at a given temperature in relation to its initial content of moisture. For most practical purposes, moisture is in the form of water vapor but the same principles apply, with different values and humidity charts, for other volatile components [22-31].

Thermal Drying consumes from 9–25% of national industrial energy consumption in the developed countries. In order to reduce net energy consumption in the drying operation there are attractive alternatives for drying of heat sensitive materials. Leather industry involves a crucial energy-intensive drying stage to remove the moisture left. Synthetic leather drying is the removal of the organic solvent and water. Determining drying characteristics for leathers is vitally important so as to optimize the drying stage. This paper describes a way to determine the drying characteristics of leather with analytical method developed for this purpose. The model presented, is based on fundamental heat and mass transfer equations. Altering air velocity varies drying conditions. The work indicates closest agreement with the theoretical model. The results from the parametric study provide a better understanding of the drying mechanisms and may lead to a series of recommendations for drying optimization. Among the many processes that are performed in the leather industry, drying has an essential role: by this means, leathers can acquire their final texture, consistency and flexibility. However, some of the unit operations involved in leather industry, especially the drying process, are still based on empiricism and tradition, with very little use of scientific principles. Widespread methods of leather drying all over the world are mostly convective methods requiring a lot of energy. Specific heat energy consumption increases, especially in the last period of the drying process, when the moisture content of the leather approaches the value at which the product is storable. However, optimizing the drying process using mathematical analysis of temperature and moisture distribution in the material can reduce energy consumption in a convective dryer. Thus, development of a suitable mathematical model to predict the accurate performance of the dryer is important for energy conservation in the drying process [32-40].

The manufacturing of new-generation synthetic leathers involves the extraction of the filling polymer from the polymer-matrix system with an organic solvent and the removal of the solvent from the highly porous material. In this paper, a mathematical model of synthetic leather drying for removing the organic solvent is proposed. The model proposed adequately describes the real processes. To improve the accuracy of calculated moisture distributions a velocity correction factor (VCF) introduced into the calculations. The VCF reflects the fact that some of the air flowing through the bed does not participate very effectively in drying, since it is channeled into low-density areas of the inhomogeneous bed. The present Chapter discusses the results of experiments to test the deductions that increased rates of drying and better agreement between predicted and experimental moisture distributions in the drying bed can be obtained by using higher air velocities.

The present work focuses on reviewing convective heat and mass transfer equations in the industrial leather drying process with particular reference to VCF [41-50].

1.2 MATERIALS AND METHODS

The theoretical model proposed in this article is based on fundamental equations to describe the simultaneous heat and mass transfer in porous media. It is possible to assume the existence of a thermodynamic quasi equilibrium state, where the temperatures of gaseous, liquid and solid phases are equal, i.e.,

$$T_S = T_L = T_G = T. \tag{1}$$

Liquid Mass Balance:

$$\frac{\partial(\varepsilon_L \rho_L)}{\partial t} + \nabla(\rho_L \vec{u}_L) + \dot{m} = 0 \tag{2}$$

Water Vapor Mass Balance:

$$\frac{\partial[(\varepsilon - \varepsilon_L)X_V \rho_G]}{\partial t} + \nabla(X_V \rho_G \vec{u}_G + \vec{J}_V) - \dot{m} = 0 \tag{3}$$

$$\vec{J}_V = -\rho_G(\varepsilon - \varepsilon_L)D_{EFF}\nabla X_V \tag{4}$$

Air Mass Balance:

$$\frac{\partial((\varepsilon - \varepsilon_L)X_A \rho_G)}{\partial t} + \nabla(X_A \rho_G \vec{u}_G - \vec{J}_V) = 0 \tag{5}$$

Liquid Momentum Eq. (Darcy's Law):

$$\vec{u}_L = -\left(\frac{\alpha_G}{\mu_G}\right)\nabla(P_G) \qquad (6)$$

Thermal Balance:

The thermal balance is governed by Eq. (7).

$$\frac{\partial\left\{\left[\rho_S C_{p_S} + (\varepsilon - \varepsilon_L)\rho_G\left(X_V C_{p_V} + X_A C_{p_A}\right) + \varepsilon_L \rho_L C_{p_L}\right]T\right\}}{\partial t} - \nabla\left(k_E \nabla T\right) +$$

$$\nabla\left[\left(\rho_L \vec{u}_L C_{p_L} + \rho_G \vec{u}_G\left(X_V C_{p_V} + X_A C_{p_A}\right)\right)T\right] + (\varepsilon - \varepsilon_L)\frac{\partial P_G}{\partial t} + \dot{m}\Delta H_V = 0 \qquad (7)$$

Thermodynamic Equilibrium-Vapor mass Fraction:

In order to attain thermal equilibrium between the liquid and vapor phase, the vapor mass fraction should be such that the partial pressure of the vapor $\left(P'_V\right)$ should be equal to its saturation pressure $\left(P_{VS}\right)$ at temperature of the mixture. Therefore, thermodynamic relations can obtain the concentration of vapor in the air/vapor mixture inside the pores. According to Dalton's Law of Additive Pressure applied to the air/vapor mixture, one can show that:

$$\rho_G = \rho_V + \rho_A \qquad (8)$$

$$X_V = \frac{\rho_V}{\rho_G} \qquad (9)$$

$$\rho_V = \frac{P'}{R_V T} \qquad (10)$$

$$\rho_A = \frac{\left(P_G - P'_V\right)}{R_A T} \qquad (11)$$

Combining Eqs. (8)–(11), one can obtain:

$$X_V = \cfrac{1}{1 + \left(\dfrac{P_G R_V}{P'_V R_A}\right) - \left(\dfrac{R_V}{R_A}\right)} \tag{12}$$

Mass Rate of Evaporation:

The mass rate of evaporation was obtained in two different ways, as follows:

First of all, the mass rate of evaporation $\dot{m}$ was expressed explicitly by taking it from the water vapor mass balance (Eq. (2)), since vapor concentration is given by Eq. (12).

$$\dot{m} = \frac{\partial[(\varepsilon - \varepsilon_L)X_V \rho_G]}{\partial t} + \nabla\left(X_V \rho_G \vec{u}_G + \vec{J}_V\right) \tag{13}$$

Secondly, an equation to compute the mass rate of evaporation can be derived with a combination of the liquid mass balance (Eq. (1)) with a first-order-Arthenius type equation. From the general kinetic equation:

$$\frac{\partial \alpha}{\partial t} = -kf(\alpha) \tag{14}$$

$$k = A \exp\left(-\frac{E}{RT_{SUR}}\right) \tag{15}$$

$$\alpha = 1 - \frac{\varepsilon_L(t)}{\varepsilon_0} \tag{16}$$

Drying Kinetics Mechanism Coupling:

Using thermodynamic relations, according to Amagat's law of additive volumes, under the same absolute pressure,

$$m_V = \frac{V_V P_G}{R_V T} \tag{17}$$

$$m_A = \frac{V_A P_G}{R_A T} \tag{18}$$

$$m_V = X_V m_T \tag{19}$$

$$m_T = m_V + m_A \tag{20}$$

$$V_G = V_V + V_A \tag{21}$$

$$V_G = (\varepsilon - \varepsilon_L) V_S \tag{22}$$

Solving the set of algebraic Eqs. (17)–(22), one can obtain the vapor-air mixture density:

$$\rho_G = \frac{(m_V + m_A)}{V_G} \tag{23}$$

$$\rho_V = \frac{m_V}{V_G} \tag{24}$$

$$\rho_A = \frac{m_A}{V_G} \tag{25}$$

Equivalent Thermal Conductivity:

It is necessary to determine the equivalent value of the thermal conductivity of the material as a whole, since no phase separation was considered in the overall energy equation. The equation we can propose now whichmay be used to achieve the equivalent thermal conductivity of materials K_E, composed of a continued medium with a uniform disperse phase. It is expressed as follows in Eq. (26).

$$K_E = \left[\frac{k_S + \varepsilon_L k_L \left(\dfrac{3k_S}{2k_S + k_L} \right) + k_G (\varepsilon - \varepsilon_L) \left(\dfrac{3k_S}{2k_S + k_G} \right)}{1 + \varepsilon_L \left(\dfrac{3k_S}{2k_S + k_L} \right) + (\varepsilon - \varepsilon_L) \left(\dfrac{3k_S}{2k_S + k_G} \right)} \right] \tag{26}$$

$$k_G = X_V k_V + X_A k_A \tag{27}$$

Effective Diffusion Coefficient Equation:

The binary bulk diffusivity D_{AV} of air-water vapor mixture is given by:

$$D_{AV} = (2.20)(10^{-5}) \left(\frac{P_{ATM}}{P_G} \right) \left(\frac{T_{REF}}{273.15} \right)^{1.75} \tag{28}$$

Factor α_F can be used to account for closed pores resulting from different nature of the solid, which would increase gas outflow resistance, so the equation of effective diffusion coefficient D_{EFF} for fiber drying is:

$$D_{EFF} = \alpha_F D_{AV} \tag{29}$$

The convective heat transfer coefficient can be expressed as:

$$h = Nu_\delta \left(\frac{k}{\delta} \right) \tag{30}$$

The convective mass transfer coefficient is:

$$h_M = \left(\frac{h}{C_{PG}}\right)\left(\frac{\mathrm{Pr}}{Sc}\right)^{2/3} \tag{31}$$

$$\mathrm{Pr} = \frac{C_{PG}\mu_G}{k_G} \tag{32}$$

$$Sc = \frac{\mu_G}{\rho_G D_{AV}} \tag{33}$$

The deriving force determining the rate of mass transfer inside the fiber is the difference between the relative humidities of the air in the pores and the fiber. The rate of moisture exchange is assumed to be proportional to the relative humidity difference in this study.

The heat transfer coefficient between external air and fibers surface can beob-

tained by: $h = Nu_\delta\left(\frac{k}{\delta}\right)$.

The mass transfer coefficient was calculated using the analogy between heat

transfer and mass transfer as $h_M = \left(\frac{h}{C_{PG}}\right)\left(\frac{\mathrm{Pr}}{Sc}\right)^{2/3}$. The convective heat and mass transfer coefficients at the surface are important parameters in drying processes; they are functions of velocity and physical properties of the drying medium.

Describing kinetic model of the moisture transfer during drying as follows:

$$-\frac{dX}{dt} = k(X - X_e) \tag{34}$$

where, X is the material moisture content (dry basis) during drying (kg water/ kg dry solids), X_e is the equilibrium moisture content of dehydrated material (kg water/kg dry solids), k is the drying rate ($\min^{-1}$), and t is the time of drying (min). The drying rate is determined as the slope of the falling rate-drying curve. At zero time, the moisture content (dry basis) of the dry material X (kg water/kg dry solids) is equal to X_i, and Eq. (34) is integrated to give the following expression:

$$X = X_e - (X_e - X_i)e^{-kt} \tag{35}$$

Using above equation Moisture Ratio can be defined as follows:

$$\frac{X - X_e}{X_i - X_e} = e^{-kt} \tag{36}$$

This is the Lewis's formula introduced in 1921. But using experimental data of leather drying it seemed that there was aerror in curve fitting of e^{-at}.

The experimental moisture content data were nondimensionlized using the equation:

$$MR = \frac{X - X_e}{X_i - X_e} \tag{37}$$

where MR is the moisture ratio. For the analysis it was assumed that the equilibrium moisture content, X_e, was equal to zero.

Selected drying models, detailed in Table 1, were fitted to the drying curves (MR versus time), and the equation parameters determined using nonlinear least squares regression analysis, as shown in Table 2.

TABLE 1   Drying models fitted to experimental data.

Model	Mathematical Expression
Lewis (1921)	$MR = \exp(-at)$
Page (1949)	$MR = \exp(-at^b)$

TABLE 1   (*Continued*)

Henderson and Pabis (1961)	$MR = a\exp(-bt)$
Quadratic function	$MR = a + bt + ct^2$
Logarithmic (Yaldiz and Eterkin, 2001)	$MR = a\exp(-bt) + c$
3rd Degree Polynomial	$MR = a + bt + ct^2 + dt^3$
Rational function	$MR = \dfrac{a + bt}{1 + ct + dt^2}$
Gaussian model	$MR = a\exp\left(\dfrac{-(t - b)^2}{2c^2}\right)$
Present model	$MR = a\exp(-bt^c) + dt^2 + et + f$

TABLE 2   Estimated values of coefficients and statistical analysis for the drying models.

Model	Constants	T = 50	T = 65	T = 80
Lewis	a	0.08498756	0.1842355	0.29379817
	S	0.0551863	0.0739872	0.0874382
	r	0.9828561	0.9717071	0.9587434
Page	a	0.033576322	0.076535988	0.14847589
	b	1.3586728	1.4803604	1.5155253
	S	0.0145866	0.0242914	0.0548030
	r	0.9988528	0.9972042	0.9856112
Henderson	a	1.1581161	1.2871764	1.4922171
	b	0.098218605	0.23327801	0.42348755
	S	0.0336756	0.0305064	0.0186881
	r	0.9938704	0.9955870	0.9983375

TABLE 2   *(Continued)*

Logarith-mic	a	1.246574	1.3051319	1.5060514
	b	0.069812589	0.1847816	0.43995186
	c	−0.15769402	−0.093918118	0.011449769
	S	0.0091395	0.0117237	0.0188223
	r	0.9995659	0.9993995	0.9985010
Quadratic function	a	1.0441166	1.1058544	1.1259588
	b	−0.068310663	−0.16107942	−0.25732004
	c	0.0011451149	0.0059365371	0.014678241
	S	0.0093261	0.0208566	0.0673518
	r	0.9995480	0.9980984	0.9806334
3rd. Degree Polynomial	a	1.065983	1.1670135	1.3629748
	b	−0.076140508	−0.20070291	−0.45309695
	c	0.0017663191	0.011932525	0.053746805
	d	−1.335923e−005	−0.0002498328	−0.0021704758
	S	0.0061268	0.0122273	0.0320439
	r	0.9998122	0.9994013	0.9961941
Rational function	a	1.0578859	1.192437	1.9302135
	b	−0.034944627	−0.083776453	−0.16891461
	c	0.03197939	0.11153663	0.72602847
	d	0.0020339684	0.01062793	0.040207428
	S	0.0074582	0.0128250	0.0105552
	r	0.9997216	0.9993413	0.9995877

TABLE 2 *(Continued)*

Gaussian model	a	1.6081505	2.3960741	268.28939
	b	−14.535231	−9.3358707	−27.36335
	c	15.612089	7.7188252	8.4574493
	S	0.0104355	0.0158495	0.0251066
	r	0.9994340	0.9989023	0.9973314
Present model	a	0.77015136	2.2899114	4.2572457
	b	0.073835826	0.58912095	1.4688178
	c	0.85093985	0.21252159	0.39672164
	d	0.00068710356	0.0035759092	0.0019698297
	e	−0.037543605	−0.094581302	−0.03351435
	f	0.3191907	−0.18402789	0.04912732
	S	0.0061386	0.0066831	0.0092957
	r	0.9998259	0.9998537	0.9997716

The experimental results for the drying of leather are given in Fig. 7. Fitting curves for two sample models (Lewis model and present model) and temperature of 80°C are given in Figs. 8 and 9. Two criteria were adopted to evaluate the goodness of fit of each model, the Correlation Coefficient (r) and the Standard Error (S). The standard error of the estimate is defined as follows:

$$S = \sqrt{\frac{\sum_{i=i}^{n_{points}} (MR_{exp,i} - MR_{Pred,i})^2}{n_{points} - n_{param}}} \tag{38}$$

where $MR_{exp,i}$ is the measured value at point , and $MR_{Pred,i}$ is the predicted value at that point, and n_{param} is the number of parameters in the particular model (so that the denominator is the number of degrees of freedom).

To explain the meaning of correlation coefficient, we must define some terms used as follow:

$$S_t = \sum_{i=1}^{n_{points}} (\bar{y} - MR_{exp,i})^2 \tag{39}$$

where, the average of the data points ($\bar{y}$) is simply given by

$$\bar{y} = \frac{1}{n_{points}} \sum_{i=1}^{n_{points}} MR_{exp,i} \tag{40}$$

The quantity S_t considers the spread around a constant line (the mean) as opposed to the spread around the regression model. This is the uncertainty of the dependent variable prior to regression. We also define the deviation from the fitting curve as:

$$S_r = \sum_{i=1}^{n_{points}} (MR_{exp,i} - MR_{pred,i})^2 \tag{41}$$

Note the similarity of this expression to the standard error of the estimate given above; this quantity likewise measures the spread of the points around the fitting function. In view of the above, the improvement (or error reduction) due to describing the data in terms of a regression model can be quantified by subtracting the two quantities. Because the magnitude of the quantity is dependent on the scale of the data, this difference is normalized to yield.

$$r = \sqrt{\frac{S_t - S_r}{S_t}} \tag{42}$$

where, is defined as the correlation coefficient. As the regression model better describes the data, the correlation coefficient will approach unity. For a perfect fit, the standard error of the estimate will approach = 0 and the correlation coefficient will approach r = 1.

The standard error and correlation coefficient values of all models are given in Figs. 10 and 11.

1.3   RESULTS AND DISCUSSION

Synthetic leathers are materials with much varied physical properties. As a consequence, even though a lot of research of simulation of drying of porous media has been carried out, the complete validation of these models are very difficult. The drying mechanisms might be strongly influenced by parameters such as permeability and effective diffusion coefficients. The unknown effective diffusion coefficient of vapor for fibers under different temperatures may be determined by adjustment of the model's theoretical alpha correction factor and experimental data. The mathematical model can be used to predict the effects of many parameters on the temperature variation of the fibers. These parameters include the operation conditions of the dryer, such as the initial moisture content of the fibers, heat and mass transfer coefficients, drying air moisture content, and dryer air temperature. From Figs. 1– 6 it can be observed that the shapes of the experimental and calculated curves are somewhat different. It can bee seen that as the actual air velocity used in this experiment increases, the value of VCF necessary to achieve reasonable correspondence between calculation and experiment becomes closer to unity; i.e., a smaller correction to air velocity is required in the calculations as the actual air velocity increases. This appears to confirm the fact that the loss in drying efficiency caused by bed inhomogeneity tends to be reduced as air velocity increases. Figure 7 shows a typical heat distribution during convective drying. Table 3 relates the VCF to the values of air velocity actuall y used in the experiments It is evident from the table that the results show a steady improvement in agreement between experiment and calculation (as indicated by increase in VCF) for air velocities up to 1.59 m/s, above which to be no further improvement with increased flow.

TABLE 3   Variation of VCF with air velocity.

Air velocity, m/s	0.75	0.89	0.95	1.59	2.10	2.59
VCF used	0.39	0.44	0.47	0.62	0.62	0.61

In this work, the analytical model has been applied to several drying experiments. The results of the experiments and corresponding calculated distributions are shown in Figs. 1- 6. It is apparent from the curves that the calculated distribution is in reasonable agreement with the corresponding experimental one. In view of the above, it can be clearly observed that the shapes of experimental and calculated curves are some what similar.

It is observed that the high air velocities will reduce the thickness of the stationary gas film on the surface of the solid and hence increase the heat and mass transfer coefficients. In practical designing of dryers it is found to be more reliable to consider heat transfer rates than mass transfer rates, as the latter are a function of surface temperature of the wet solid, which is difficult to determine and cannot, in practice, be assumed to be that of the wet-bulb temperature of the air with an adequate degree of accuracy [51-108].

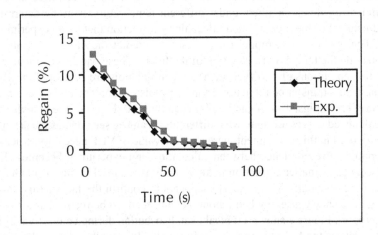

FIGURE 1   Comparison of the theoretical and experimental distribution at air velocity of 0.75 m/s and VCF = 0.39.

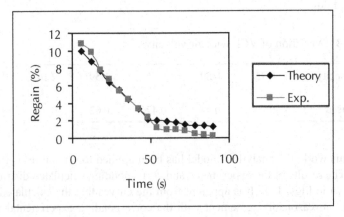

FIGURE 2   Comparison of the theoretical and experimental distribution at air velocity of 0.89 m/s and VCF = 0.44.

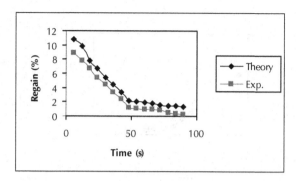

FIGURE 3   Comparison of the theoretical and experimental distribution at air velocity of 0.95 m/s and VCF = 0.47.

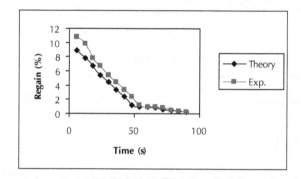

FIGURE 4   Comparison of the theoretical and experimental Distribution at air velocity of 1.59 m/s and VCF = 0.62.

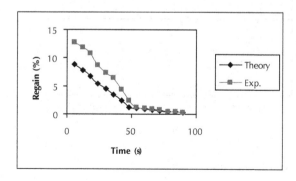

FIGURE 5   Comparison of the theoretical and experimental distribution at air velocity of 2.10 m/s and VCF = 0.62.

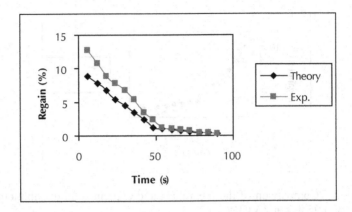

FIGURE 6 Comparison of the theoretical and experimental distribution at air velocity of 2.59. m/s and VCF = 0.61.

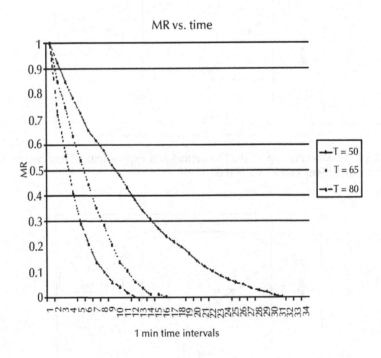

FIGURE 7 Moisture Ratio vs. Time.

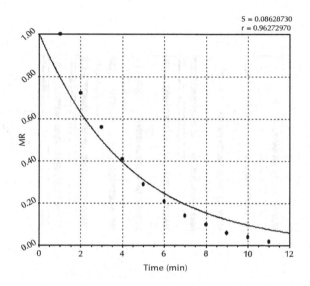

FIGURE 8   Lewis model.

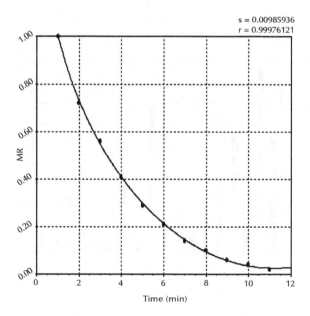

FIGURE 9   Present model.

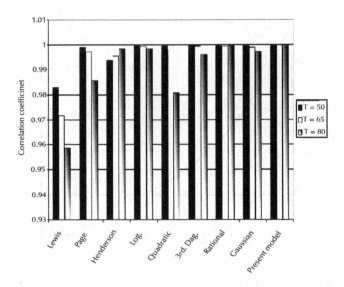

FIGURE 10   Correlation coefficient of all models.

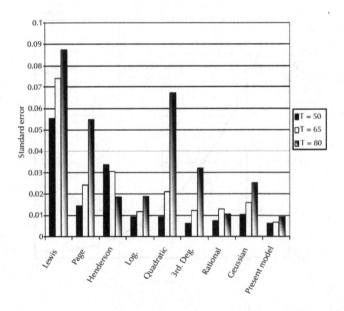

FIGURE 11   Standard error of all models.

1.4 CONCLUSIONS

In the model presented in this book, a simple method of predicting moisture distributions leads to prediction of drying times more rapid than those measured in experiments. From this point of view, the drying reveals many aspects, which are not normally observed or measured, and which may be of value in some application.

The derivation of the drying curves is an example. It is clear from the experiments over a range of air velocities that it is not possible to make accurate predictions and have the experimental curves coincide at all points with the predicted distributions simply by introducing a VCF into the calculations. This suggest that a close agreement between calculated and experimental curves over the entire drying period could be obtained by using a large value of VCF in the initial stages of drying and progressively decreasing it as drying proceeds.

KEYWORDS

- **Statistical analysis**
- **Air velocity**
- **Moisture content**
- **Drying models**
- **Kinetic model**

REFERENCES

1. Joukowski, N. Paper to Polytechnic Soc. Moscow, spring of 1898m English translation by Miss O.; *Simin. Proc. AWWA*, **1904**, 57–58.
2. Allievi, L. "General Theory of Pressure Variation in Pipes,"*Ann. D. Ing.* **1982**, 166–171.
3. Wood, F. M."History of Water hammer," Civil Engineering Research Report, #65, Queens University, Canada, **1970**, 66–70.
4. Parmakian J."Water hammer Design Criteria,"*J. Power Div. ASCE*, Sept.; **1957**, 456–460.
5. Parmakian J."Water hammer Analysis," Dover Publications, Inc.: New York, **1963**, 51–58.

6. Streeter V. L.; Lai C.; "Water hammer Analysis Including Fluid Friction." *J. Hydraulics Division, ASCE,* **1962,** *88,* 79 p.

7. Streeter V. L.; Wylie, E. B.; "Fluid Mechanics," McGraw-Hill Ltd.: USA, **1979, 492–505.**

8. Streeter V. L.; Wylie, E. B.; "Fluid Mechanics," McGraw-Hill Ltd.: USA, **1981, 398–420.**

9. Wylie, E. B.; Streeter V. L.; Talpin, L. B.; Matched impedance to control fluid transients. *Trans. ASME* **1983,** *105(2),* 219–224.

10. Wylie, E. B.; Streeter V. L.; Fluid Transients in Systems, Prentice-Hall, Englewood Cliffs, New Jersey, **1993,** 4 p.

11. Wylie, E. B.; Streeter V. L.; Fluid Transients, Feb Press, Ann Arbor, MI, **1982,** corrected copy, **1983,** 158 p.

12. Brunone B.; Karney, B. W.; Mecarelli M.; Ferrante M. "Velocity Profiles and Unsteady Pipe Friction in Transient Flow" Journal of Water Resources Planning and Management, ASCE, **2000,** *126(4),* Jul.; 236–244.

13. Koelle E.; Luvizotto Jr. E.; Andrade, J. P. G.; "Personality Investigation of Hydraulic Networks using MOC – Method of Characteristics" Proceedings of the 7th International Conference on Pressure Surges and Fluid Transients, Harrogate Durham, United Kingdom, **1996,** 1–8.

14. Filion Y.; Karney, B. W.; "A Numerical Exploration of Transient Decay Mechanisms in Water Distribution Systems," Proceedings of the ASCE Environmental Water Resources Institute Conference, American Society of Civil Engineers, Roanoke, Virginia, **2002,** 30.

15. Hamam, M. A.; Mc Corquodale, J. A.; "Transient Conditions in the Transition from Gravity to Surcharged Sewer Flow,"*Canadian J. Civil Eng.,* Canada, Sep.; **1982,** 65–98.

16. Savic, D. A.; Walters, G. A.; "Genetic Algorithms Techniques for Calibrating Network Models," Report No. 95/12, Centre for Systems and Control Engineering, **1995,** 137–146.

17. Savic, D. A.; Walters, G. A.; Genetic Algorithms Techniques for Calibrating Network Models, University of Exeter, Exeter, United Kingdom, **1995,** 41–77.

18. Walski, T. M.; Lutes, T. L.; "Hydraulic Transients Cause Low-Pressure Problems," Journal of the American Water Works Association, **1994,** *75(2),* 58.

19. Wu, Z. Y.; Simpson, A. R.; Competent genetic-evolutionary optimization of water distribution systems. *J. Comput. Civ. Eng.***2001,** *15(2),* 89–101.

20. Gerasimov Yu. I.; The course of physical chemistry. V. 1. Goskhimizdat, M.; **1963,** 736.

21. Dikarevsky M.; Impactprotection opositelnyh closed systems. Moscow: Kolos, **1981,** 80.

22. Nigmatulin, R. I.; Nagiyev, F. B.; Khabeev, N. S.; Destruction and collapse of vapor bubbles and strengthening shock waves in a liquid with vapor bubbles. Assembly. "Gas and wave dynamics," No.3,"MSU," **1979,** 124–129.

23. Allievi L.; "General Theory of Pressure Variation in Pipes,"*Ann. D. Ing.* **1982,** 166–171.

24. Joukowski N.; Paper to Polytechnic Soc. Moscow, spring of1898, English translation by Miss O.; Simin. Proc. AWWA, **1904,** 57–58.

25. Wisniewski, K. P.; Design of pumping stations closed irrigation systems: Right. Vishnevsky, K. P.; Podlasov, A. V.; Moscow: Agropromizdat, **1990,** 93.
26. Nigmatulin, R. I.; Khabeev, N. S.; Nagiyev, F. B.; Dynamics, heat and mass transfer of vapor-gas bubbles in a liquid. *Int.J. Heat Mass Transfer,* **1981,** *24(6),* Printed in Great Britain, 1033–1044.
27. Vargaftik, N. B.; Handbook of thermo-physical properties of gases and liquids. Oxford: Pergamon Press, **1972,** 98.
28. Laman, B. F.; Hydro pumps and installation, **1988,** 278.
29. Nagiyev, F. B.; Kadyrov, B. A.; Heat transfer and the dynamics of vapor bubbles in a liquid binary solution. DAN Azerbaijani, S. S. R.; **1986,** *4,* 10–13.
30. Alyshev V. M.; Hydraulic calculations of open channels on your PC. – Part 1 Tutorial. – Moscow: MSUE, **2003,** 185.
31. Streeter V. L.; Wylie, E. B.; "Fluid Mechanics," McGraw-Hill Ltd.; USA, **1979,** 492–505.
32. Sharp B.; "Water hammer Problems and Solutions," Edward Arnold Ltd.; London, **1981,** 43–55.
33. Skousen P.; "Valve Handbook," McGraw Hill, New York, HAMMER Theory and Practice, **1998,** 687–721.
34. Shaking, N. I.; Water hammer to break the continuity of the flow in pressure conduits pumping stations: Dis. on Kharkov, **1988,** 225.
35. Tijsseling,"Alan E Vardy Time scales and FSI in unsteady liquid-filled pipe flow," **1993,** 5–12.
36. Wu, P. Y.; Little W. A.; Measurement of friction factor for flow of gases in very fine channels used for micro miniature, Joule Thompson refrigerators, *Cryogenics* 24 (8), **1983,** 273–277.
37. Song C. C. et al.; "Transient Mixed-Flow Models for Storm Sewers,"*J. Hyd. Div.* Nov.; **1983,** *109,* 458–530.
38. Stephenson D.; "Pipe Flow Analysis," Elsevier, *19,* S. A.; **1984,** 670–788.
39. Chaudhry, M. H.; "Applied Hydraulic Transients," Van Nostrand Reinhold Co.; N. Y.; **1979,** 1322–1324.
40. Chaudhry, M. H.; Yevjevich V. "Closed Conduit Flow," Water Resources Publication, USA, **1981,** 255–278.
41. Chaudhry, M. H.; Applied Hydraulic Transients, Van Nostrand Reinhold: New York, USA, **1987,** p.165–167.
42. Kerr, S. L.; "Minimizing service interruptions due to transmission line failures: Discussion," Journal of the American Water Works Association, *41, 634,* July **1949,** 266–268.
43. Kerr, S. L.; "Water hammer control," Journal of the American Water Works Association, **1951,** *43,* December985–999.
44. Apoloniusz Kodura, Katarzyna Weinerowska," Some Aspects of Physical and Numerical Modeling of Water Hammer in Pipelines," **2005,** 126–132.
45. Anuchina, N. N.; Volkov V. I.; Gordeychuk V. A.; Es'kov, N. S.; Ilyutina, O. S.; Kozyrev O. M. "Numerical simulations of Rayleigh-Taylor and Richtmyer-Meshkov instability using mah-3 code,"*J. Comput. Appl. Math.***2004,** *168,* 11.
46. Fox, J. A.; "Hydraulic Analysis of Unsteady Flow in Pipe Network," Wiley, N. Y.; **1977,** 78–89.

47. Karassik, I. J.; "Pump Handbook – Third Edition," McGraw-Hill, **2001,** 19–22.
48. Fok, A.; "Design Charts for Air Chamber on Pump Pipelines,"*J. Hyd. Div. ASCE,* Sept.; **1978,** 15–74.
49. Fok, A.; Ashamalla A.; Aldworth G.; "Considerations in Optimizing Air Chamber for Pumping Plants," Symposium on Fluid Transients and Acoustics in the Power Industry, San Francisco, USA, Dec, **1978,** 112–114.
50. Fok, A.; "Design Charts for Surge Tanks on Pump Discharge Lines," BHRA 3rd Int. Conference on Pressure Surges, Bedford, England, Mar.; **1980,** 23–34.
51. Fok, A.; "Water hammer and Its Protection in Pumping Systems," Hydro technical Conference, CSCE, Edmonton, May, **1982,** 45–55.
52. Fok, A.; "A contribution to the Analysis of Energy Losses in Transient Pipe Flow," PhD Thesis, University of Ottawa, **1987,** 176–182.
53. Hariri Asli, K.; Nagiyev, F. B.; Water Hammer and fluid condition, Ministry of Energy, Gilan Water and Wastewater Co.; Research Week Exhibition, Tehran, Iran, December, **2007,** 132–148, http://isrc.nww.co.ir.
54. Hariri Asli, K.; Nagiyev, F. B.; Water Hammer analysis and formulation, Ministry of Energy, Gilan Water and Wastewater Co.; Research Week Exhibition, Tehran, Iran, December, **2007,** 111–131, http://isrc.nww.co.ir.
55. Hariri Asli, K.; Nagiyev, F. B.; Water Hammer and hydrodynamics instabilities, Interpenetration of two fluids at parallel between plates and turbulent moving in pipe, Ministry of Energy, Guilan Water and Wastewater Co.; Research Week Exhibition, Tehran, Iran, December, **2007,** 90–110, http://isrc.nww.co.ir.
56. Hariri Asli, K.; Nagiyev, F. B.; Water Hammer and pump pulsation, Ministry of Energy, Guilan Water and Wastewater Co.; Research Week Exhibition, Tehran, Iran, December, **2007,** 51–72, http://isrc.nww.co.ir.
57. Hariri Asli, K.; Nagiyev, F. B.; Reynolds number and hydrodynamics' instability," Ministry of Energy, Guilan Water and Wastewater Co.; Research Week Exhibition, Tehran, Iran, December, **2007,** 31–50, http://isrc.nww.co.ir.
58. Hariri Asli, K.; Nagiyev, F. B.; Water Hammer and valves, Ministry of Energy, Guilan Water and Wastewater Co.; Research Week Exhibition, Tehran, Iran, December, **2007,** 20–30, http://isrc.nww.co.ir.
59. Hariri Asli, K.; Nagiyev, F. B.; "Interpenetration of two fluids at parallel between plates and turbulent moving in pipe," Ministry of Energy, Guilan Water and Wastewater Co.; Research Week Exhibition, Tehran, Iran, December, **2007,** 73–89, http://isrc.nww.co.ir.
60. Hariri Asli, K.; Nagiyev, F. B.; Decreasing of Unaccounted For Water "UFW" by Geographic Information System"GIS" in Rasht urban water system, civil engineering organization of Guilan, Technical and Art Journal, **2007,** 3–7, http://www.art-of-music.net/.
61. Hariri Asli, K.; Portable Flow meter Tester Machine Apparatus, Certificate on registration of invention, Tehran, Iran, #010757, Series a/82, 24/11/2007, 1–3
62. Hariri Asli, K.; Nagiyev, F. B.; Haghi, A. K.; "Interpenetration of two fluids at parallel between plates and turbulent moving in pipe," 9th Conference on Ministry of Energetic works at research week, Tehran, Iran, **2008,** 73–89, http://isrc.nww.co.ir.

63. Hariri Asli, K.; Nagiyev, F. B.; Haghi, A. K.; "Water hammer and valves," 9th Conference on Ministry of Energetic works at research week, Tehran, Iran, **2008**, 20–30, http://isrc.nww.co.ir.
64. Hariri Asli, K.; Nagiyev, F. B.; Haghi, A. K.; "Water hammer and hydrodynamics instability," 9th Conference on Ministry of Energetic works at research week, Tehran, Iran, **2008**, 90–110, http://isrc.nww.co.ir.
65. Hariri Asli, K.; Nagiyev, F. B.; Haghi, A. K.; "Water hammer analysis and formulation," 9th Conference on Ministry of Energetic works at research week, Tehran, Iran, **2008**, 27–42, http://isrc.nww.co.ir.
66. Hariri Asli, K.; Nagiyev, F. B.; Haghi, A. K.; "Water hammer &fluid condition," 9th Conference on Ministry of Energetic works at research week, Tehran, Iran, **2008**, 27–43, http://isrc.nww.co.ir.
67. Hariri Asli, K.; Nagiyev, F. B.; Haghi, A. K.; "Water hammer and pump pulsation," 9th Conference on Ministry of Energetic works at research week, Tehran, Iran, **2008**, 27–44, http://isrc.nww.co.ir.
68. Hariri Asli, K.; Nagiyev, F. B.; Haghi, A. K.; "Reynolds number and hydrodynamics instability," 9th Conference on Ministry of Energetic works at research week, Tehran, Iran, **2008**, 27–45, http://isrc.nww.co.ir.
69. Hariri Asli, K.; Nagiyev, F. B.; Haghi, A. K.; "Water hammer and fluid Interpenetration," 9th Conference on Ministry of Energetic works at research week, Tehran, Iran, **2008**, 27–47, http://isrc.nww.co.ir.
70. Hariri Asli, K.; GIS and water hammer disaster at earthquake in Rasht water pipeline, civil engineering organization of Guilan, Technical and Art Journal, **2008**, 14–17, http://www.art-of-music.net/.
71. Hariri Asli, K.; GIS and water hammer disaster at earthquake in Rasht water pipeline, 3rd International Conference on Integrated Natural Disaster Management, Tehran university, ISSN: 1735–5540, 18–19 Feb.; INDM, Tehran, Iran, **2008**, *13*, 53/1–12, http://www.civilica.com/Paper-INDM03-INDM03_001.html
72. Hariri Asli, K.; Nagiyev, F. B.; Bubbles characteristics and convective effects in the binary mixtures. Transactions issue mathematics and mechanics series of physical-technical and mathematics science, ISSN: 0002–3108, Azerbaijan, Baku, **2009**, 68–74, http://www.imm.science.az/journals.html.
73. Hariri Asli, K.; Nagiyev, F. B.; Haghi, A. K.; Aliyev, S. A.; Three-Dimensional conjugate heat transfer in porous media, 1st Festival on Water and Wastewater Research and Technology, Tehran, Iran, 12–17 Dec. **2009**, 26–28, http://isrc.nww.co.ir.
74. Hariri Asli, K.; Nagiyev, F. B.; Haghi, A. K.; Aliyev, S. A.; Some Aspects of Physical and Numerical Modeling of water hammer in pipelines, 1st Festival on Water and Wastewater Research and Technology, Tehran, Iran, 12–17 Dec. **2009**, 26–29, http://isrc.nww.co.ir
75. Hariri Asli, K.; Nagiyev, F. B.; Haghi, A. K.; Aliyev, S. A.; Modeling for Water Hammer due to valves: From theory to practice, 1st Festival on Water and Wastewater Research and Technology, Tehran, Iran, 12–17 Dec. **2009**, 26–30, http://isrc.nww.co.ir.
76. Hariri Asli, K.; Nagiyev, F. B.; Haghi, A. K.; Aliyev, S. A.; Water hammer and hydrodynamics instabilities modeling: From Theory to Practice, 1st Festival on Water and

Wastewater Research and Technology, Tehran, Iran, 12–17 Dec. **2009**, 26–31, http://isrc.nww.co.ir

77. Hariri Asli, K.; Nagiyev, F. B.; Haghi, A. K.; Aliyev, S. A.; A computational approach to study fluid movement, 1st Festival on Water and Wastewater Research and Technology, Tehran, Iran, 12–17 Dec. **2009**, 27–32, http://isrc.nww.co.ir.

78. Hariri Asli, K.; Nagiyev, F. B.; Haghi, A. K.; Aliyev, S. A.; Water Hammer Analysis: Some Computational Aspects and practical hints, 1st Festival on Water and Wastewater Research and Technology, Tehran, Iran, 12–17 Dec. **2009**, 27–33, http://isrc.nww.co.ir

79. Hariri Asli, K.; Nagiyev, F. B.; Haghi, A. K.; Aliyev, S. A.; Water Hammer and Fluid condition: A computational approach, 1st Festival on Water and Wastewater Research and Technology, Tehran, Iran, 12–17 Dec.; 2009, 27–34, http://isrc.nww.co.ir.

80. Hariri Asli, K.; Nagiyev, F. B.; Haghi, A. K.; Aliyev, S. A.; A computational Method to Study Transient Flow in Binary Mixtures, 1st Festival on Water and Wastewater Research and Technology, Tehran, Iran, 12–17 Dec. **2009**, 27–35, http://isrc.nww.co.ir.

81. Hariri Asli, K.; Nagiyev, F. B.; Haghi, A. K.; Physical modeling of fluid movement in pipelines, 1st Festival on Water and Wastewater Research and Technology, Tehran, Iran, 12–17 Dec. **2009**, 27–36, http://isrc.nww.co.ir.

82. Hariri Asli, K.; Nagiyev, F. B.; Haghi, A. K.; Aliyev, S. A.; Interpenetration of two fluids at parallel between plates and turbulent moving, 1st Festival on Water and Wastewater Research and Technology, Tehran, Iran, 12–17 Dec. **2009**, 27–37, http://isrc.nww.co.ir.

83. Hariri Asli, K.; Nagiyev, F. B.; Haghi, A. K.; Aliyev, S. A.; Modeling of fluid interaction produced by water hammer, 1st Festival on Water and Wastewater Research and Technology, Tehran, Iran, 12–17 Dec.2009, 27–38, http://isrc.nww.co.ir.

84. Hariri Asli, K.; Nagiyev, F. B.; Haghi, A. K.; Aliyev, S. A.; GIS and water hammer disaster at earthquake in Rasht pipeline, 1st Festival on Water and Wastewater Research and Technology, Tehran, Iran, 12–17 Dec. **2009**, 27–39, http://isrc.nww.co.ir.

85. Hariri Asli, K.; Nagiyev, F. B.; Haghi, A. K.; Aliyev, S. A.; Interpenetration of two fluids at parallel between plates and turbulent moving, 1st Festival on Water and Wastewater Research and Technology, Tehran, Iran, 12–17 Dec. **2009**, 27–40, http://isrc.nww.co.ir.

86. Hariri Asli, K.; Nagiyev, F. B.; Haghi, A. K.; Aliyev, S. A.; Water hammer and hydrodynamics' instability, 1st Festival on Water and Wastewater Research and Technology, Tehran, Iran, 12–17 Dec. **2009**, 27–41, http://isrc.nww.co.ir.

87. Hariri Asli, K.; Nagiyev, F. B.; Haghi, A. K.; Aliyev, S. A.; Water hammer analysis and formulation, 1st Festival on Water and Wastewater Research and Technology, Tehran, Iran, 12–17 Dec. **2009**, 27–42, http://isrc.nww.co.ir.

88. Hariri Asli, K.; Nagiyev, F. B.; Haghi, A. K.; Aliyev, S. A.; Water hammer & fluid condition, 1st Festival on Water and Wastewater Research and Technology, Tehran, Iran, 12–17 Dec. **2009**, 27–43, http://isrc.nww.co.ir.

89. Hariri Asli, K.; Nagiyev, F. B.; Haghi, A. K.; Aliyev, S. A.; Water hammer and pump pulsation, 1st Festival on Water and Wastewater Research and Technology, Tehran, Iran, 12–17 Dec. **2009**, 27–44, http://isrc.nww.co.ir.

90. Hariri Asli, K.; Nagiyev, F. B.; Haghi, A. K.; Aliyev, S. A.; Reynolds number and hydrodynamics instabilities, 1st Festival on Water and Wastewater Research and Technology, Tehran, Iran, 12–17 Dec. **2009**, 27–45, http://isrc.nww.co.ir.

91. Hariri Asli, K.; Nagiyev, F. B.; Haghi, A. K.; Aliyev, S. A.; "Water Hammer and Valves,"1st Festival on Water and Wastewater Research and Technology, Tehran, Iran, 12–17 Dec. **2009**, 27–46, http://isrc.nww.co.ir.

92. Hariri Asli, K.; Nagiyev, F. B.; Haghi, A. K.; Aliyev, S. A.; "Water Hammer and Fluid Interpenetration,"1st Festival on Water and Wastewater Research and Technology, Tehran, Iran, 12–17 Dec. **2009**, 27–47, http://isrc.nww.co.ir.

93. Hariri Asli, K.; Nagiyev, F. B.; Modeling of fluid interaction produced by water hammer, International Journal of Chemoinformatics and Chemical Engineering, IGI, ISSN: 2155–4110, EISSN: 2155–4129, USA, **2010**, 29–41, http://www.igi-global.com/journals/details.asp?ID=34654

94. Hariri Asli, K.; Nagiyev, F. B.; Haghi, A. K.; Water hammer and fluid condition; a computational approach, Computational Methods in Applied Science and Engineering, USA, Chapter 5, Nova Science Publications, ISBN: 978-1-60876-052-7, USA, **2010**, 73–94, https://www.novapublishers.com/catalog/

95. Hariri Asli, K.; Nagiyev, F. B.; Haghi, A. K.; Some aspects of physical and numerical modeling of water hammer in pipelines. Computational Methods in Applied Science and Engineering, USA, Chapter 23, Nova Science Publications, ISBN: 978-1-60876-052-7, USA, **2010**, 365–387, https://www.novapublishers.com/catalog/

96. Hariri Asli, K.; Nagiyev, F. B.; Haghi, A. K.; Modeling for water hammer due to valves; from theory to practice, Computational Methods in Applied Science and Engineering, USA, Chapter 11, Nova Science Publications ISBN: 978-1-60876-052-7, USA, **2010**, 229–236, https://www.novapublishers.com/catalog/

97. Hariri Asli, K.; Nagiyev, F. B.; Haghi, A. K.; A computational method to Study transient flow in binary mixtures, Computational Methods in Applied Science and Engineering, USA, Chapter 13, Nova Science Publications ISBN: 978-1-60876-052-7, USA, **2010**, 229–236, https://www.novapublishers.com/catalog/

98. Hariri Asli, K.; Nagiyev, F. B.; Haghi, A. K.; Water hammer analysis; some computational aspects and practical hints, Computational Methods in Applied Science and Engineering, USA, Chapter 16, Nova Science Publications ISBN: 978-1-60876-052-7, USA, **2010**, 263–281, https://www.novapublishers.com/catalog/

99. Hariri Asli, K.; Nagiyev, F. B.; Haghi, A. K.; Water hammer and hydrodynamics instabilities modeling, Computational Methods in Applied Science and Engineering, USA, Chapter 17, From Theory to Practice, Nova Science Publications ISBN: 978-1-60876-052-7, USA, **2010**, 283–301, https://www.novapublishers.com/catalog/

100. Hariri Asli, K.; Nagiyev, F. B.; Haghi, A. K.; A computational approach to study water hammer and pump pulsation phenomena, Computational Methods in Applied Science and Engineering, USA, Chapter 22, Nova Science Publications, ISBN: 978-1-60876-052-7, USA, **2010**, 349–363, https://www.novapublishers.com/catalog/

101. Hariri Asli, K.; Nagiyev, F. B.; Haghi, A. K.; A computational approach to study fluid movement, Nanomaterials Yearbook –**2009**, From Nanostructures, Nanomaterials and Nanotechnologies to Nanoindustry, Chapter 16, Nova Science Publications, USA, ISBN: 978-1-60876-451-8, USA, **2010**, 181–196, https://www.novapublishers.com/catalog/product_info.php?products_id=11587

102. Hariri Asli, K.; Nagiyev, F. B.; Haghi, A. K.; Physical modeling of fluid movement in pipelines, Nanomaterials Yearbook –2009, From Nanostructures, Nanomaterials and Nanotechnologies to Nanoindustry, Chapter 17, Nova Science Publications, USA, ISBN: 978-1-60876-451-8, USA, 2010, 197–214, https://www.novapublishers.com/catalog/product_info.php?products_id =11587

103. Hariri Asli, K.; Nagiyev, F. B.; Haghi, A. K.; "Some Aspects of Physical and Numerical Modeling of water hammer in pipelines," Nonlinear Dynamics An International Journal of Nonlinear Dynamics and Chaos in Engineering Systems, ISSN: 1573–269X (electronic version) Journal no. 11071 Springer, Netherlands, 2009, ISSN: 0924–090X (print version), Springer, Heidelberg, Germany, Number 4 / June, 2010, Volume 60, 677–701, http://www.springerlink.com/openurl.aspgenre=article&id=doi: 10.1007/s11071-009-9624-7.

104. Hariri Asli, K.; Nagiyev, F. B.; Haghi, A. K.; Interpenetration of two fluids at parallel between plates and turbulent moving in pipe; a case study, Computational Methods in Applied Science and Engineering, USA, Chapter 7, Nova Science Publications, ISBN: 978-1-60876-052-7, USA, 2010, 107–133, https://www.novapublishers.com/catalog/

105. Hariri Asli, K.; Nagiyev, F. B.; Beglou, M. J.; Haghi, A. K.; Kinetic analysis of convective drying, International Journal of the Balkan Tribological Association, ISSN: 1310–4772, Sofia, Bulgaria, 2009, 15(4), 546–556, jbalkta@gmail.com

106. Hariri Asli, K.; Nagiyev, F. B.; Haghi, A. K.; Three-dimensional Conjugate Heat Transfer in Porous Media, International Journal of the Balkan Tribological Association, ISSN: 1310–4772, Sofia, Bulgaria, 2009, 15(3), 336–346, jbalkta@gmail.com

107. Hariri Asli, K.; Nagiyev, F. B.; Haghi, A. K.; Aliyev, S. A.; Pure Oxygen penetration in wastewater flow, Recent Progress in Research in Chemistry and Chemical Engineering, Nova Science Publications, ISBN: 978-1-61668-501-0, Nova Science Publications, USA, 2010, 17–27, https://www.novapublishers.com/catalog/product_info.php?products_id=13174110100.

108. Hariri Asli, K.; Nagiyev, F. B.; Haghi, A. K.; Aliyev, S. A.; Improved modeling for prediction of water transmission failure, Recent Progress in Research in Chemistry and Chemical Engineering, Nova Science Publications, ISBN: 978-1-61668-501-0, Nova Science Publications, USA, 2010, 28–36, https://www.novapublishers.com/catalog/product_info.php?products_id=13174.

CHAPTER 2

DISPERSED FLUID AND IDEAL FLUID MECHANICS

CONTENTS

2.1 INTRODUCTION

In this book, miscible liquids condition, for example, velocity–pressure–temperature and the other properties is as similar and the main approach is the changes study on behavior of the fluids flow state. According to Reynolds number magnitude (RE. NO.), separation of fluid direction happened. For fluid motion modeling, 2D-component disperses fluid motion used. Modeling of two-phase liquid–liquid flows through a Kinetics static mixer by means of computational fluid dynamics (CFD) has been presented. The two-modeled phases were assumed viscous and Newtonian with the physical properties mimicking an aqueous solution in the continuous and oil in the dispersed (secondary) phase. Differential equations included in the proposed model describe the unsteady motion of a real fluid through the channels and pipes. These differential equations are derived from the following assumptions. It was assumed that the pipe is cylindrical with a constant cross-sectional area with the initial pressure. The fluid flow through the pipe is the one-dimensional. It is assumed that the characteristics of resistors, fixed for steady flows and unsteady flows are equivalent.

One of the problems in the study of fluid flow in plumbing systems is the behavior of stratified fluid in the channels. Mostly steady flows initially are ideal, then the viscous and turbulent fluid in the pipes [1-9] .

2.2 MATERIALS AND METHODS

A fluid flow is compressible if its density ρ changes appreciably within the domain of interest. Typically, this will occur when the fluid velocity exceeds Mach 0.3. Hence, low velocity flows (both gas and liquids) behave incompressibly. An incompressible fluid is one whose density is constant everywhere. All fluids behave incompressibly (to within 5%) when their maximum velocities are below Mach 0.3. Mach number is the relative velocity of a fluid compared to its sonic velocity. Mach numbers less than 1 correspond to subsonic velocities, and Mach numbers > 1 corresponds to super-sonic velocities. A Newtonian fluid [1-34] is a viscous fluid whose shear stresses is a linear function of the fluid strain rate. Mathematically, this can be expressed as: $\tau_{ij} = K_{ijqp} \times D_{pq}$, where τ_{ij} is the shear stress component, and D_{pq} are fluid strain rate components [10-12]

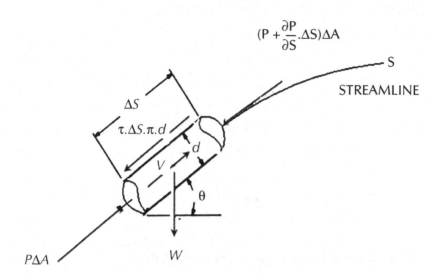

FIGURE 1   Newton second law (conservation of momentum equation) for fluid element.

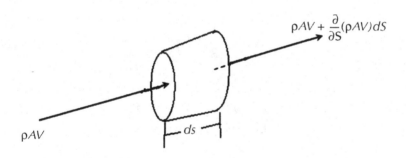

FIGURE 2   Continuity equation (conservation of mass) for fluid element.

It is defined as the combination of momentum equation (Fig.1) and continuity equation (Fig.2) for determining the velocity and pressure in a one-dimensional flow system. The solving of these equations produces a theoretical result that usually corresponds quite closely to actual system measurements.

$$P\Delta A - (P + \frac{\partial P}{\partial S}.\Delta S)\Delta A - W.\sin\theta - \tau.\Delta S.\pi.d = \frac{W}{g}.\frac{dV}{dt} \qquad (1)$$

Both sides are divided by m and with assumption:

$$\frac{\partial Z}{\partial S} = +\sin\theta,$$

(2)

$$-\frac{1}{\partial}\cdot\frac{\partial P}{\partial S} - \frac{\partial Z}{\partial S} - \frac{4\tau}{\gamma D} = \frac{1}{g}\cdot\frac{dV}{dt},$$

(3)

$$\Delta A = \frac{\Pi.D^2}{4},$$

(4)

If fluid diameter assumed equal to pipe diameter, then:

$$\frac{-1}{\gamma}\cdot\frac{\partial P}{\partial S} - \frac{\partial Z}{\partial S} - \frac{4\tau_\circ}{\gamma.D},$$

(5)

$$\tau_\circ = \frac{1}{8}\rho.f.V^2,$$

(6)

$$\frac{1}{\gamma}\cdot\frac{\partial P}{\partial S} - \frac{\partial Z}{\partial S} - \frac{f}{D}\cdot\frac{V^2}{2g} = \frac{1}{g}\cdot\frac{dV}{dt},$$

(7)

$$V^2 = V\,|V|,\ \frac{1}{\gamma}\cdot\frac{\partial P}{\partial S} - \frac{\partial Z}{\partial S} - \frac{f}{D}\cdot\frac{V^2}{2g} = \frac{1}{g}\cdot\frac{dV}{dt},$$

(8)

(Euler equation)

For finding (V) and (P) we need to "conservation of mass law" (Fig.2):

$$\rho AV - \left[\rho AV - \frac{\partial}{\partial S}(\rho AV)dS\right] = \frac{\partial}{\partial t}(\rho AdS) - \frac{\partial}{\partial S}(\rho AV)dS = \frac{\partial}{\partial t}(\rho AdS)$$

(9)

$$-\left(\rho A \frac{\partial V}{\partial S}dS + \rho V \frac{\partial A}{\partial S}dS + AV \frac{\partial \rho}{\partial S}dS\right) = \rho A \frac{\partial}{\partial t}(dS) + \rho dS \frac{\partial A}{\partial t} + AdS \frac{\partial p}{\partial t},$$

(10)

$$\frac{1}{\rho}\left(\frac{\partial \rho}{\partial t} + V \frac{\partial \rho}{\partial S}\right) + \frac{1}{A}\left(\frac{\partial A}{\partial t} + V \frac{\partial A}{\partial S}\right) + \frac{1}{dS} \cdot \frac{\partial}{\partial t}(dS) + \frac{\partial V}{\partial S} = \circ$$

With $\dfrac{\partial \rho}{\partial t} + V \dfrac{\partial \rho}{\partial S} = \dfrac{d\rho}{dt}$ and $\dfrac{\partial A}{\partial t} + V \dfrac{\partial A}{\partial S} = \dfrac{dA}{dt}$

$$\frac{1}{\rho} \cdot \frac{d\rho}{dt} + \frac{1}{A} \cdot \frac{dA}{dt} + \frac{\partial V}{\partial S} + \frac{1}{dS} \cdot \frac{1}{dt}(dS) = \circ,$$

(11)

$$K = \left(\frac{d\rho}{\left(\dfrac{d\rho}{\rho}\right)}\right)$$

(Fluid module of elasticity) then:

$$\frac{1}{\rho} \cdot \frac{d\rho}{dt} = \frac{1}{k} \cdot \frac{d\rho}{dt},$$

(12)

Put Eq. (7) into Eq. (8) Then:

$$\frac{\partial V}{\partial S} + \frac{1}{k} \cdot \frac{d\rho}{dt} + \frac{1}{A} \cdot \frac{dA}{dt} + \frac{1}{dS} \cdot \frac{d}{dt}(dS) = \circ,$$

(13)

$$\rho\frac{\partial V}{\partial S}+\frac{d\rho}{dt}\rho\left[\frac{1}{k}+\frac{1}{A}\cdot\frac{dA}{d\rho}+\frac{1}{dS}\cdot\frac{d}{d\rho}(dS)\right]=\circ, \qquad (14),$$

$$K=\left(\frac{d\rho}{\left(\frac{d\rho}{\rho}\right)}\right)\text{(Fluid module of elasticity)}, \qquad (15)$$

$$\rho\left[\frac{1}{k}+\frac{1}{A}\cdot\frac{dA}{dt}+\frac{1}{dS}\cdot\frac{d}{d\rho}(dS)\right]=\frac{1}{C^2}, \qquad (16)$$

Then
$$C^2\frac{\partial V}{\partial S}+\frac{1}{\rho}\cdot\frac{d\rho}{dt}=\circ, \qquad (17)$$

(Continuity equation)

Partial differential Eqs.(4) and (10) are solved by method of characteristics "MOC":

$$\frac{dp}{dt}=\frac{\partial p}{\partial t}+\frac{\partial p}{\partial S}\cdot\frac{dS}{dt}, \qquad (18)$$

$$\frac{dV}{dt}=\frac{\partial V}{\partial t}+\frac{\partial V}{\partial S}\cdot\frac{dS}{dt}, \qquad (19)$$

Then,

$$\left|\frac{\partial V}{\partial t}+\frac{1}{\rho}\frac{\partial p}{\partial S}+g\frac{dz}{dS}+\frac{f}{2D}V|V|=\circ,\right.$$
$$\left|C^2\frac{\partial V}{\partial S}+\frac{1}{\rho}\frac{\partial P}{\partial t}=\circ,\right. \qquad (20)$$

By Linear combination of Eqs. (13) and (14)

$$\lambda\left(\frac{\partial V}{\partial t}+\frac{1}{\rho}\frac{\partial p}{\partial S}+g\cdot\frac{dz}{dS}+\frac{f}{2D}V|V|\right)+C^2\frac{\partial V}{\partial S}+\frac{1}{\rho}\frac{\partial p}{\partial t}=\circ, \qquad (21)$$

$$\left(\lambda\frac{\partial V}{\partial t}+C^2\frac{\partial V}{\partial S}\right)+\left(\frac{1}{\rho}\cdot\frac{\partial\rho}{\partial t}+\frac{\lambda}{\rho}\cdot\frac{\partial P}{\partial S}\right)+\lambda.g.\frac{dz}{dS}+\frac{\lambda.f}{2D}V|V|=\circ\,, \quad (22)$$

$$\lambda\frac{\partial V}{\partial t}+C^2\frac{\partial V}{\partial S}=\lambda\frac{dV}{dt}\Rightarrow\lambda\frac{dS}{dt}=C^2\,, \quad (23)$$

$$\frac{1}{\rho}\cdot\frac{\partial p}{\partial t}+\frac{\lambda}{\rho}\cdot\frac{\partial\rho}{\partial S}=\frac{1}{\rho}\cdot\frac{d\rho}{dt}\Rightarrow$$
$$\frac{\lambda}{\rho}=\frac{1}{\rho}\cdot\frac{dS}{dt} \quad , \quad (24)$$

$$\left|\frac{C^2}{\lambda}=\lambda\text{ (By removing }\frac{dS}{dt}\text{)}, \quad \lambda=\pm C\right.$$

For $\lambda=\pm C$, from Eq. (18) we have:

$$\frac{dV}{dt}+\frac{1}{\rho}\cdot\frac{dp}{dt}+C.g.\frac{dz}{dS}+C.\frac{f}{2D}V|V|=\circ\,, \quad (25)$$

Dividing both sides by "C" we get:

$$\frac{dV}{dt}+\frac{1}{c.\rho}\frac{dP}{dt}+g.\frac{dz}{dS}+\frac{f}{2D}V|V|=\circ\,, \quad (26)$$

For $\lambda=-C$ by Eq. (16):

$$\frac{dV}{dt}+\frac{1}{c.\rho}\frac{dP}{dt}+g.\frac{dz}{dS}+\frac{f}{2D}V|V|=\circ\,, \quad (27)$$

If $\rho = \rho.g(H - Z)$,   (28)

From Eqs. (9) and (10):

$$\left| \begin{array}{l} \dfrac{dV}{dt} + \dfrac{g}{c} \cdot \dfrac{dH}{dt} + \dfrac{f}{2D} V|V| = \circ \\[2mm] if : \dfrac{dS}{dt} = C, \end{array} \right.$$ (29)

$$\left| \begin{array}{l} \dfrac{dV}{dt} + \dfrac{g}{c} \cdot \dfrac{dH}{dt} + \dfrac{f}{2D} V|V| = \circ, \\[2mm] if : \dfrac{dS}{dt} = -C, \end{array} \right.$$ (30)

The method of characteristics is a finite difference technique which pressures (Figs.3 and 4) were computed along the pipe for each time step (1)–(35).

Calculation automatically subdivided the pipe into sections (intervals) and selected a time interval for computations Eqs. (22) and (24) are the characteristic equation of Eqs. 21 and 23.

If, $f = 0$; Then, Eq. (23) will be (Figs.3 and 4):

$$\frac{dV}{dt} - \frac{g}{c} \cdot \frac{dH}{dt} = \circ$$

or

$$dH = \left(\frac{C}{g} \right) dV, (Zhukousky),$$ (31)

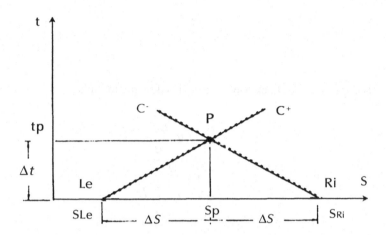

FIGURE 3   Intersection of characteristic lines with positive and negative slope.

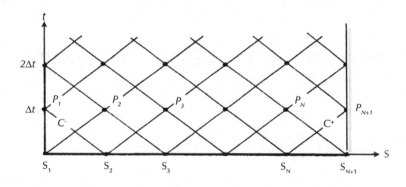

FIGURE 4   Set of characteristic lines intersection for assumed pipe by finite difference method of water.

If the pressure at the inlet of the pipe and along its length is equal to P_0, then slugging pressure undergoes a sharp increase:

$$\Delta p : p = p_0 + \Delta p ,$$

$$(32)$$

The Zhukousky formula is as flowing:

$$\Delta p = \left(\frac{C.\Delta V}{g} \right),$$

(33)

The speed of the shock wave is calculated by the formula:

$$C = \sqrt{\frac{g. \dfrac{E_W}{\rho}}{1 + \dfrac{d}{t_W} \cdot \dfrac{E_W}{E}}},$$

(34)

Hammer: $T_p - 0 = \Delta t$

$$c^{+} : (Vp - VLe)/(T_P - \circ) + \left(\frac{g}{c} \right)(Hp - HLe)/(TP - o) + fV_{Le}|V_{Le}|/2D) = \circ,$$

(35)

$$c^{-} : (Vp - VRi)/(Tp - 0) + \left(\frac{g}{c} \right)\left(Hp - HRi\right)/(TP - o) + fV_{Ri}|V_{Ri}|/2D) = \circ,$$

(36)

$$c^{+} : (Vp - VLe) + \left(\frac{g}{c} \right)(Hp - HLe) + (f.\Delta t)(f.V_{Le}|V_{Le}|/2D) = \circ,$$

(37)

$$c^{-} : (Vp - VRi) + \left(\frac{g}{c} \right)(Hp - HRi) + (f.\Delta t)(fV_{Ri}|V_{Ri}|/2D) = \circ,$$

(38)

$$V_p = \frac{1}{2}\left[(V_{Le}+V_{Ri}) + \frac{g}{c}\left(H_{Le}-H_{Ri}\right) - (f.\Delta t/2D)(V_{Le}|V_{Le}| - VR_i|V_{Ri}|)\right],$$

(39)

$$H_p = \frac{1}{2}\left[\frac{c}{g}(V_{Le}+V_{Ri}) + (H_{Le}-H_{Ri}) - \frac{c}{g}(f.\Delta t/2D)(V_{Le}|V_{Le}| - V_{Ri}|V_{Ri}|)\right],$$

(40)

$V_{Le}, V_{Ri}, H_{Le}, HRi, f, D$ are initial conditions parameters.

They are applied for solution at steady state condition. Water hammer equations calculation starts with pipe length "L" divided by "N" parts:

$$\Delta S = \frac{L}{N} \ \& \ \Delta t = \frac{\Delta s}{C},$$

(41)

Eqs. (28) and (29) are solved for the range P_2 through P_N, therefore H and V are found for internal points. Therefore:

At P_1 there is only one characteristic Line (c^-)

At P_{N+1} there is only one characteristic Line (c^+)

For finding H and V at P_1 and P_{N+1} the boundary conditions are used.

The Lagrangian approach was used to track the trajectory of dispersed fluid elements (drops) in the simulated static mixer. The particle history was analyzed in terms of the residence time in the mixer. While two relaxing miscible fluids (35-50) are mixed together, their appearances in terms of colors and shapes will change due to their mixing interpenetration (Fig. 5).

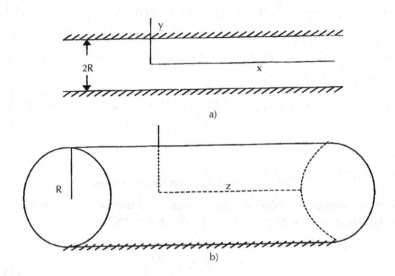

FIGURE 5   Two Dimensional fluids flow.

Use equations of motion of two relaxing fluids in pipe are as flowing:

$$u_1 = u_1(y,t) \ , \ \ u_2 = u_2(y,t)$$

$$\left. \begin{array}{l} \rho_1 \dfrac{\partial u_1}{\partial t} = f_1 \mu_1 \dfrac{\partial^2 u_1}{\partial y^2} + k(u_2 - u_1) - f_1 \dfrac{\partial p}{\partial x} \ , \\[3mm] \rho_2 \dfrac{\partial u_2}{\partial t} = f_2 \mu_2 \dfrac{\partial^2 u_2}{\partial y^2} + k(u_1 - u_2) - f_2 \dfrac{\partial p}{\partial x} \ , \\[3mm] \dfrac{\partial p}{\partial y} = 0 \ , \ \ \dfrac{\partial p}{\partial z} = 0 \ , \ \ f_1 + f_2 = 0 \end{array} \right\} \qquad (42)$$

$\overline{u}$, – velocit y (m/s), p – pressure, k – module of elasticity of water (kg/m²), f – Darcy-Weisbach friction factor (obtained from Moody diagram) for each pipe, μ – fluid dynamic, viscosity (kg/m.s), ρ – density (kg/m³).

Calculation for equation of motion for relaxing fluids:

$$\left.\begin{array}{l} \theta_1 \dfrac{\partial \tau_1}{\partial t} + \tau_1 = \mu_1 \dfrac{\partial u_1}{\partial y}, \\[4mm] \theta_2 \dfrac{\partial \tau_2}{\partial t} + \tau_2 = \mu_2 \dfrac{\partial u_2}{\partial y} \end{array}\right\} \tag{43}$$

θ_1, θ_2 – relaxing time of fluids, define equation of motion for Interpenetration of two 2D pressurized relaxing fluids at parallel between plates and turbulent moving in pipe as flowing:

$$\left.\begin{array}{l} \rho_1 \dfrac{\partial u_1}{\partial t} + \rho_1 \theta_1 \dfrac{\partial^2 u_1}{\partial t^2} = f_1 \mu_1 \dfrac{\partial^2 u_1}{\partial y^2} + \theta_1 k \dfrac{\partial(u_1 - u_2)}{\partial t} + k(u_2 - u_1) - f_1 \left[\theta_1 \dfrac{\partial^2 p}{\partial t \partial x} + \dfrac{\partial p}{\partial x} \right] \\[4mm] \rho_2 \dfrac{\partial u_2}{\partial t} + \rho_2 \theta_2 \dfrac{\partial^2 u_2}{\partial t^2} = f_2 \mu_2 \dfrac{\partial^2 u_2}{\partial y^2} + \theta_2 k \dfrac{\partial(u_1 - u_2)}{\partial t} + k(u_1 - u_2) - f_2 \left[\theta_2 \dfrac{\partial^2 p}{\partial t \partial x} + \dfrac{\partial p}{\partial x} \right] \\[4mm] \dfrac{\partial p}{\partial y} = 0, \quad \dfrac{\partial p}{\partial z} = 0, \\[4mm] f_1 + f_2 = 0 \end{array}\right\}$$

$$\tag{1.3}$$

From Eq. (3) concluded that pressure drop $\partial p / \partial x$ it is not effective but time is effective.

Assumed that at first time both plan are stopped and pressure at coordination for this time is low.

$$\left.\begin{array}{l} t = 0 \left\{ \begin{array}{l} u_1 = 0, \ u_2 = 0 \\ \partial u_1 / \partial t = 0, \ \partial u_2 / \partial t = 0 \end{array} \right. \\[4mm] y = h \quad (t > 0) \quad u_1 = 0 \quad u_2 = 0 \\[2mm] y = -h \quad (t > 0) \quad u_1 = 0 \quad u_2 = 0 \end{array}\right\}, \tag{44}$$

At time t condition with Laplace rule, with Eqs.(3) and (4) we have:

$$\left.\begin{array}{l} \dfrac{d^2\bar{u}_1}{dy^2} - \alpha_1\bar{u}_1 + \beta_1\bar{u}_2 = \dfrac{1}{\mu_1}\dfrac{\partial P}{\partial x} \\[3mm] \dfrac{d^2\bar{u}_2}{dy^2} - \alpha_2\bar{u}_2 + \beta_2\bar{u}_1 = \dfrac{1}{\mu_2}\dfrac{\partial P}{\partial x} \end{array}\right\} , \qquad (45)$$

With:

$$\left.\begin{array}{l} y = h \quad \bar{u}_1 = 0 \ , \ \bar{u}_2 = 0 \\[2mm] y = -h \quad \bar{u}_1 = 0 \ , \ \bar{u}_2 = 0 \end{array}\right\} , \qquad (46)$$

Where,

$$\left.\begin{array}{l} \alpha_1 = \dfrac{\rho_1(\theta_1 s^2 + s) + k(\theta_1 s + 1)}{f_1\mu_1} \ , \\[4mm] \beta_1 = \dfrac{k(\theta_1 s + 1)}{f_1\mu_1} \ , \\[4mm] \alpha_2 = \dfrac{\rho_2(\theta_2 s^2 + s) + k(\theta_2 s + 1)}{f_2\mu_2} \ , \\[4mm] \beta_2 = \dfrac{k(\theta_2 s + 1)}{f_2\mu_2} \ , \end{array}\right\} \qquad (47)$$

Calculation $\partial p / \partial x = A = const$ and with product of Eq. (5) into N flowing differential equation received:

$$\frac{d^2}{dy^2}(N\bar{u}_1 + \bar{u}_2) - (\alpha_2 - N\beta_1)(N\bar{u}_1 + \bar{u}_2) = \left[\frac{N(1+\theta_1 s)}{\mu_1} + \frac{1+\theta_2 s}{\mu_2}\right] A \ , \ (48)$$

$$N_{1,2} = \frac{-(\alpha_1 - \alpha_2) \pm \sqrt{(\alpha_1 - \alpha_2)^2 + 4\beta_1\beta_2}}{2\beta_1} . \qquad (49)$$

Eq. (48) calculated with Eq. (49):

$$A\overline{u}_1 + \overline{u}_2 = -A\left[\frac{N}{\mu_1} + \frac{1}{\mu_2}\right]\left[1 - \frac{ch\sqrt{\alpha_2 - N\beta_1}y}{ch\sqrt{\alpha_2 - N\beta_1}h}\right].$$ (50)

N calculation with two meaning:

$$N_1\overline{u}_1 + \overline{u}_2 = -A\left[\frac{N_1}{\mu_1} + \frac{1}{\mu_2}\right]\left[1 - \frac{ch\sqrt{\alpha_2 - N_1\beta_1}y}{ch\sqrt{\alpha_2 - N_1\beta_1}h}\right],$$ (51)

$$N_2\overline{u}_1 + \overline{u}_2 = -A\left[\frac{N_2}{\mu_1} + \frac{1}{\mu_2}\right]\left[1 - \frac{ch\sqrt{\alpha_2 - N_2\beta_1}y}{ch\sqrt{\alpha_2 - N_2\beta_1}h}\right],$$ (52)

Where for equation velocity find:

$$\overline{u}_1 = \frac{A}{N_2 - N_1}\left\{\frac{\frac{N_1}{\mu_1} + \frac{1}{\mu_2}}{\alpha_2 - N_1\beta_1}\left[1 - \frac{ch\sqrt{\alpha_2 - N_1\beta_1}y}{ch\sqrt{\alpha_2 - N_1\beta_1}h}\right] - \frac{\frac{N_2}{\mu_1} + \frac{1}{\mu_2}}{\alpha_2 - N_2\beta_1}\left[1 - \frac{ch\sqrt{\alpha_2 - N_2\beta_1}y}{ch\sqrt{\alpha_2 - N_2\beta_1}h}\right]\right\},$$

$$\overline{u}_2 = \frac{A}{N_1 - N_2}\left\{\frac{\frac{N_2}{\mu_1} - \frac{\beta_2}{\beta_1}\frac{1}{\mu_2}}{\alpha_2 - N_1\beta_1}\left[1 - \frac{ch\sqrt{\alpha_2 - N_1\beta_1}y}{ch\sqrt{\alpha_2 - N_1\beta_1}h}\right] - \frac{\frac{N_1}{\mu_2} - \frac{\beta_2}{\beta_1}\frac{\mu_1}{\mu_2}}{\alpha_2 - N_2\beta_1}\left[1 - \frac{ch\sqrt{\alpha_2 - N_2\beta_1}y}{ch\sqrt{\alpha_2 - N_2\beta_1}h}\right]\right\}.$$

$$\overline{\mu}_i = \frac{1}{2\pi}\int_{\sigma+i\infty}^{\sigma+i\infty}\frac{A}{N_2 - N_1}\left\{\frac{\frac{N_1}{\mu_1} - \frac{1}{\mu_2}}{\alpha_2 - N_1\beta_1}\left[1 - \frac{ch\sqrt{\alpha_2 - N_1\beta_1}y}{ch\sqrt{\alpha_2 - N_1\beta_1}h}\right] - \right.$$

$$\left. -\frac{\dfrac{N_2}{\mu_1}-\dfrac{1}{\mu_2}}{\alpha_2-N_2\beta_1}\left[1-\frac{ch\sqrt{\alpha_2-N_2\beta_1}y}{ch\sqrt{\alpha_2-N_2\beta_1}h}\right]\right\}.e^{st}\frac{ds}{s}, \tag{53}$$

In this calculation we have:

$$s=s_1 \quad , \quad s=s_2 \quad , \quad s=s_3 \quad , \quad s=s_4 : s_1,s_2,s_3,s_4,$$

$$S_{1n}=\gamma_{1n},S_{2n}=\gamma_{2n},S_{3n}=\gamma_{3n},S_{4n}=\gamma_{4n},\gamma_{in}$$

Proportional to forth procedure:

$$\alpha_2-N_1\beta_1=-\frac{\pi^2}{h^2}\left(n+\frac{1}{2}\right)^2, \tag{54}$$

$$\alpha_2-N_2\beta_1=-\frac{\pi^2}{h^2}\left(n+\frac{1}{2}\right)^2, \tag{55}$$

In this state for velocity we have:

$$\mu_1=-\frac{\dfrac{1}{f_1\mu_1f_2\mu_2}}{\dfrac{1}{f_1\mu_1}+\dfrac{1}{f_2\mu_2}}A\left\{-\frac{1}{2}\left(y^2-h^2\right)+\frac{\left(\dfrac{1}{\mu_1}-\dfrac{1}{\mu_2}\right)f_2\mu_2}{\left(\dfrac{1}{f_1\mu_1}+\dfrac{1}{f_2\mu_2}\right)k}\times\right.$$

$$\times\left[1-\frac{ch\left(\sqrt{\dfrac{1}{f_1\mu_1}+\dfrac{1}{f_2\mu_2}}\,ky\right)}{ch\left(\sqrt{\dfrac{1}{f_1\mu_1}+\dfrac{1}{f_2\mu_2}}\,kh\right)}\right]+\frac{4A}{\pi}\frac{4}{\,}\sum_{i=1}^{\infty}\sum_{n=1}^{\infty}\frac{(-1)^{n+1}}{\left(n+\dfrac{1}{2}\right)}cos\left[\pi\left(n+\frac{1}{2}\right)\frac{y}{h}\right]\times$$

$$\times\left\{\frac{\left[\dfrac{\pi^2}{h^2}\left(n+\dfrac{1}{2}\right)+\dfrac{\left(\theta^2\gamma_{in}+1\right)\left(\rho^2\gamma_{in}+k\right)}{f_2\mu_2}\right]\dfrac{1}{\mu_1}+\dfrac{k\left(\theta_1\gamma_{in}+1\right)}{f_1\mu_1}\dfrac{1}{\mu_2}}{\dfrac{\left(\theta_1\gamma_{in}+1\right)\left(\pi_1\gamma_{in}+k\right)}{f_1\mu_1}+\dfrac{\left(\theta_2\gamma_{in}+1\right)\left(\pi_2\gamma_{in}+k\right)}{f_2\mu_2}-2\dfrac{\pi^2}{h^2}\left(n+\dfrac{1}{2}\right)^2}\cdot\frac{1}{\gamma_{in}}\right.$$

$$\times\frac{1}{2\gamma_{in}\left(\dfrac{\theta_1\rho_1}{f_1\mu_1}+\dfrac{\theta_2\rho_2}{f_2\mu_2}\right)+\dfrac{\theta_1k_1+\rho_1}{f_1\mu_1}+\dfrac{\theta_2k+\rho_2}{f_2\mu_2}+\dfrac{\dfrac{\left(\theta_1\gamma_{in}+1\right)\left(\rho_1\gamma_{in}+k\right)}{f_1^i\,_1}}{2\dfrac{\pi^2}{h^2}\left(n+\dfrac{1}{2}\right)^2+\dfrac{\left(\theta_1\gamma_{in}+1\right)\left(\rho_1\gamma_{in}+k\right)}{f_1\mu_1}}-}$$

$$\left.\frac{e^{-\gamma t}\,_{in}}{-\dfrac{\left(\theta_2\gamma_{in}+1\right)\left(\rho_2\gamma_{in}+k\right)}{f_2\mu_2}\right)\left(2\gamma_{in}\left(\dfrac{\theta_1\rho_1}{f_1\mu_1}+\dfrac{\theta_2\rho_2}{f_2\mu_2}\right)-\dfrac{\theta_2k+\rho_2}{f_2\mu_2}-\dfrac{\theta_1k+\rho_1}{f_1\mu_1}\right)}{\left(\dfrac{\left(\theta_2\gamma_{in}+1\right)\left(\rho_2\gamma_{in}+k\right)}{f_2\mu_2}\right)}\right\},\quad(56)$$

$$u_2=-\frac{\dfrac{1}{f_1\mu_1f_2\mu_2}}{\dfrac{1}{f_1\mu_1}+\dfrac{1}{f_2\mu_2}}A\left\{-\frac{1}{2}\left(y^2+h^2\right)+\frac{\left(\dfrac{1}{\mu_1}-\dfrac{1}{\mu_2}\right)f_1\mu_1}{\left(\dfrac{1}{f_1\mu_1}-\dfrac{1}{f_2\mu_2}\right)k}\times\right.$$

$$\times\left[1-\frac{ch\sqrt{\left(\dfrac{1}{f_1\mu_1}+\dfrac{1}{f_2\mu_2}\right)}ky}{ch\sqrt{\left(\dfrac{1}{f_1\mu_1}+\dfrac{1}{f_2\mu_2}\right)}kh}\right]\Bigg\}+\frac{4A}{\pi}\sum_{i=1}^{4}\sum_{n=1}^{\infty}\frac{(-1)^{n+1}}{\left(n+\dfrac{1}{2}\right)}\cos\left[\pi\left(n+\frac{1}{2}\right)\frac{y}{h}\right]\times$$

$$\times\left\{\frac{\left[\dfrac{\pi^2}{h^2}\left(n+\dfrac{1}{2}\right)+\dfrac{(\theta^2\gamma_{in}+1)(\rho_1\gamma_{in}+k)}{f_1\mu_1}\right]\dfrac{1}{\mu_2}+\dfrac{k(\theta_1\gamma_{in}+1)}{f_2\mu_2}\dfrac{1}{\mu_1}}{\dfrac{(\theta_1\gamma_{in}+1)(\rho_1\gamma_{in}+k)}{f_1\,1}+\dfrac{(\theta_2\gamma_{in}+1)(\rho_2\gamma_{in}+k)}{f\,i_2\,2}-2\dfrac{\pi^2}{h^2}\left(n\pm\dfrac{1}{2}\right)^2}\cdot\dfrac{1}{\gamma_{in}}}{2\gamma_{in}\left(\dfrac{\theta_1\rho_1}{f_1\mu_1}+\dfrac{\theta_2\rho_2}{f_2\mu_2}\right)+\dfrac{\theta_1 k+\rho_1}{f_1\mu_1}+\dfrac{\theta_2 k+\rho_2}{f_2\mu_2}+\dfrac{\left(\dfrac{(\theta_1\gamma_{in}+1)(\rho_1\gamma_{in}+k)}{f_1\,1}\right)}{2\dfrac{\pi^2}{h^2}\left(n+\dfrac{1}{2}\right)^2+\dfrac{(\theta_1\gamma_{in}+1)(\rho_1\gamma_{in}+k)}{f_1\mu_1}}}\right.$$

$$\left.\frac{e-\gamma_{in}^t}{-\dfrac{(\theta_2\gamma_{in}+1)(\rho_2\gamma_{in}+k)}{f_2\mu_2}\Bigg]\Bigg[2\gamma_{in}\left(\dfrac{\theta_1\rho_1}{f_1\mu_1}+\dfrac{\theta_2\rho_2}{f_2\mu_2}\right)-\dfrac{\theta_2 k+\rho_2}{f_2\mu_2}-\dfrac{\theta_1 k+\rho_1}{f_1\mu_1}+\dfrac{4k^2(\theta_1\theta_2\gamma_{in}+\theta_1+\theta_z)}{f_1\mu_1 f_2\mu_2}\right]}{+\dfrac{(\theta_1\gamma_{in}+1)(\rho_1\gamma_{in}+k)}{f_1\mu_1}-\dfrac{(\theta_2\gamma_{in}+1)(\rho_2\gamma_{in}+k)}{f_2\mu_2}}\right\},(57)$$

When $\theta_1=\theta_2=0$ from Eqs. (9) and (10) we have:
$\theta_1=\theta_2=0$

$$\mu_1 = \mu_2 \quad , \quad \rho_{1i} = \rho_{2i} \; ,$$

$$u = u_1 = u_2 = \frac{A}{2\mu}\left(h^2 - y^2\right) - \frac{16h^2 A}{\pi\mu} \sum_{n=1}^{\infty} \frac{(-1)^n}{(2n+1)^3} \cos\frac{(2n+1)}{2h} y.e - \frac{\pi^2}{h^2}\left(n+\frac{1}{2}\right)^2 \frac{\mu}{\rho} t$$

At condition $t \to \infty$ for unsteady motion of fluid, it is easy for calculation table pr ocedure;

$$\left.\begin{aligned} \rho_1 \frac{\partial u_1}{\partial t} &= f_1\mu_1\left(\frac{\partial^2 u_1}{\partial r^2} + \frac{1}{r}\frac{\partial u_1}{\partial r}\right) + k(u_2 - u_1) - f_1\frac{\partial P}{\partial z} \\ \rho_2 \frac{\partial u_2}{\partial t} &= f_2\mu_2\left(\frac{\partial^2 u_2}{\partial r^2} + \frac{1}{r}\frac{\partial u_2}{\partial r}\right) + k(u_1 - u_2) - f_2\frac{\partial P}{\partial z} \end{aligned}\right\}, \qquad (58)$$

For every relaxing phase we have:

$$\left.\begin{aligned} \theta_1 \frac{\partial \tau_1}{\partial t} + \tau_1 &= \mu_1 \frac{\partial u_1}{\partial r} \; , \\ \theta_2 \frac{\partial \tau_2}{\partial t} + \tau_2 &= \mu_2 \frac{\partial u_2}{\partial r} \; , \end{aligned}\right\} \qquad (59)$$

Start and limiting conditions:

$$\begin{aligned} t = 0 \qquad & u_1 = 0 \, , u_2 = 0 \, , \\ r = R(t > 0) \qquad & u_1 = 0 \, , u_2 = 0 \, . \end{aligned} \qquad (60)$$

In condition of differential Eq. (11) by $\partial\tau_1/\partial t$ from Eq. (12) and with τ_1 concluded:

$$\left.\begin{aligned}
\rho_1\left(\frac{\partial u_1}{\partial t}+\theta_1\frac{\partial^2 u_1}{\partial t^2}\right) &= f_1\mu_1\left(\frac{\partial^2 u_1}{\partial t^2}+\frac{1}{r}\frac{\partial u_1}{\partial r}\right)+k\left[\theta_1\frac{\partial}{\partial t}(u_2-u_1)+(u_2-u_1)\right]- \\
&\quad -f_1\left[\theta_1\theta_1\frac{\partial^2 p}{\partial t\partial z}+\frac{\partial p}{\partial z}\right], \\
\rho_2\left(\frac{\partial u_2}{\partial t}+\theta_2\frac{\partial^2 u_2}{\partial t^2}\right) &= f_2\mu_2\left(\frac{\partial^2 u_2}{\partial r^2}+\frac{1}{r}\frac{\partial u_2}{\partial r}\right)+k\left[\theta_2\frac{\partial}{\partial t}(u_1-u_2)+(u_1-u_2)\right]- \\
&\quad -f_2\left[\theta_2\frac{\partial^2 p}{\partial t\partial z}+\frac{\partial p}{\partial z}\right].
\end{aligned}\right\} \quad (61)$$

Data condition Eq. (13) and integration. In this condition Laplace is toward Eq. (14). Then solution find in the form of velocity equation, *1D* fluid viscosity in round pipe is:

$$u_1 = -\frac{A}{f_1\mu_1\left(\frac{1}{f_1\mu_1}+\frac{1}{f_2\mu_2}\right)}\left\{\frac{1}{4}\left(r^2-R^2\right)\frac{1}{f_2\mu_2}+\frac{\left(\frac{1}{\mu_2}-\frac{1}{\mu_1}\right)f_2\mu_2}{k\left(\frac{1}{f_1\mu_1}+\frac{1}{f_2\mu_2}\right)}\times\right.$$

$$\times\left[1-\frac{I_0\left(\sqrt{\left(\frac{1}{f_1\mu_1}+\frac{1}{f_2\mu_2}\right)}kr\right)}{I_0\left(\sqrt{\left(\frac{1}{f_1\mu_1}+\frac{1}{f_2\mu_2}\right)}kR\right)}\right]\right\}+\sum_{i=1}^{4}\sum_{n=1}^{\infty}\frac{4A}{\alpha_n}\frac{J_0\left(\alpha_n\frac{r}{R}\right)}{J_1(\alpha_n)}\times$$

$$\times\left\{\frac{\frac{\alpha_n^2}{R^2}+\frac{(\theta_2\gamma_{in}+1)(-\rho_2\gamma_{in}+k)}{f_1\mu_1}\frac{1}{\mu_1}+\frac{k(\theta_1\gamma_{in}+1)}{f_1\mu_1}\frac{1}{\mu_2}}{\frac{(\theta_1\gamma_{in}+1)(-\rho_1\gamma_{in}+k)}{f_1\mu_1}+\frac{(\theta_2\gamma_{in}+1)(-\rho_2\gamma_{in}+k)}{f_2\mu_2}\pm2\frac{\alpha_n^2}{R^2}}\times\right.$$

$$\times \frac{e^{-\gamma_{in} t}}{\gamma_{in}} \bigg/ \left\{ 2\gamma_{in}\left(\frac{\theta_1\rho_1}{f_1\mu_1} + \frac{\theta_2\rho_2}{f_2\mu_2}\right) \pm \frac{\theta_1 k + \rho_1}{f_1\mu_1} + \frac{\theta_2 k + \rho_2}{f_2\mu_2} + \right.$$

$$+ \left[\left(\frac{(\theta_1\gamma_{in}+1)(\rho_1\gamma_{in}+k)}{f_1\mu_1} - \frac{(\theta_2\gamma_{in}+1)(\rho_2\gamma_{in}+k)}{f_2\mu_2} \right) \left(2\gamma_{in}\left(\frac{\theta_1\rho_1}{f_1\mu_1} + \frac{\theta_2\rho_2}{f_2\mu_2}\right) - \right. \right.$$

$$\left. - \frac{\theta_2 k + \rho_2}{f_2\mu_2} - \frac{\theta_1 k + \rho_1}{f_1\mu_1} + 4k^2\frac{\theta_1\theta_2\gamma_{in} + \theta_1 + \theta_2}{f_1\mu_1 f_2\mu_2} \right) \right] \bigg/ \left[2\frac{\alpha_n^2}{R} + \right.$$

$$\left. + \frac{(\theta_1\gamma_{in}+1)(-\rho_1\gamma_{in}+k)}{f_1\mu_1} + \frac{(\theta_2\gamma_{in}+1)(-\rho_2\gamma_{in}+k)}{f_2\mu_2} \right] \bigg] \bigg\}, \qquad (62)$$

$$u_2 = -\frac{A}{f_1\mu_1\left(\frac{1}{f_1\mu_1} + \frac{1}{f_2\mu_2}\right)}\left\{ \frac{1}{4}\left(r^2 - R^2\right)\frac{1}{f_2\mu_2} + \frac{\left(\frac{1}{\mu_2} - \frac{1}{\mu_1}\right)f_2\mu_2}{k\left(\frac{1}{f_1\mu_1} + \frac{1}{f_2\mu_2}\right)} \times \right.$$

$$\times \left[1 - \frac{I_0\left(\sqrt{\left(\frac{1}{f_1\mu_1} + \frac{1}{f_2\mu_2}\right)}kr\right)}{I_0\left(\sqrt{\left(\frac{1}{f_1\mu_1} + \frac{1}{f_2\mu_2}\right)}kr\right)} \right] \bigg\} + \sum_{i=1}^{4}\sum_{n=1}^{\infty}\frac{4A}{\alpha_n}\frac{J_0\left(\alpha_n\frac{r}{R}\right)}{J_1(\alpha_n)} \times$$

$$\times \left\{ \frac{\dfrac{\alpha_n^2}{R^2} + \dfrac{(\theta_2\gamma_{in}+1)(\rho_1\gamma_{in}+k)}{f_1\mu_1}\dfrac{1}{\mu_2} + \dfrac{k(\theta_2\gamma_{in}+1)}{f_2\mu_2}\dfrac{1}{\mu_1}}{\dfrac{(\theta_1\gamma_{in}+1)(-\rho_1\gamma_{in}+k)}{f_1\mu_1} + \dfrac{(\theta_2\gamma_{in}+1)(-\rho_2\gamma_{in}+k)}{f_2\mu_2} \pm 2\dfrac{\alpha_n^2}{R^2}} \right\} \times$$

$$\times \frac{e^{-\gamma_{int}}}{\gamma_{in}} \left/ \left\{ 2\gamma_{in}\left(\frac{\theta_1\rho_1}{f_1\mu_1} + \frac{\theta_2\rho_2}{f_2\mu_2}\right) + \frac{\theta_1k+\rho_1}{f_1\mu_1} + \frac{\theta_2k-\rho_2}{f_2\mu_2} \pm \right. \right.$$

$$\pm \left[\left(\frac{(\theta_1\gamma_{in}+1)(\rho_1\gamma_{in}+k)}{f_1\mu_1} - \frac{(\theta_2\gamma_{in}+1)}{f_2\mu_2}\right)\left(2\gamma_{in}\left(\frac{\theta_1\rho_1}{f_1\mu_1} + \frac{\theta_2\rho_2}{f_2\mu_2}\right) + \right.\right.$$

$$+\frac{\theta_2k+\rho_2}{f_2\mu_2} - \frac{\theta_1k+\rho_1}{f_1\mu_1} + 4k^2\frac{\theta_1\theta_2\gamma_{in}+\theta_1+\theta_2}{f_1\mu_1f_2\mu_2}\right)\right] \left/ \left[-2\frac{\alpha_n^2}{R} + \right.\right.$$

$$\left. +\frac{(\theta_1\gamma_{in}+1)(\rho_1\gamma_{in}+k)}{f_1\mu_1} - \frac{(\theta_2\gamma_{in}+1)(\rho_2\gamma_{in}+k)}{f_2\mu_2}\right]\right\}, \qquad (63)$$

When $\theta_1 = \theta_2 = 0$ from Eqs. (15) and (16) we have Eq. (4) in condition:

$$\begin{cases} \theta_1 = \theta_2 = 0 \\ \mu_1 = \mu_2 = \mu \\ \rho_{1i} = \rho_{2i} = \rho \end{cases}$$

One of the problems in the study of fluid flow in plumbing systems is the behavior of stratified fluid in the channels. Mostly steady flows initially are ideal, then the viscous and turbulent fluid in the pipes.

At the deep pool filled with water, and on its surface to create a disturbance, then the surface of the water will begin to propagate. Their origin is explained by the fact that the fluid particles are located near the cavity.

The fluid particles create disturbance, which will seek to fill the cavity under the influence of gravity. The development of this phenomenon is led to the spread of waves on the water. The fluid particles in such a wave do not move up and down around in circles. The waves of water are neither longitudinal nor transverse. They seem to be a mixture of both. The radius of the circles varies with depth of moving fluid particles. They reduce to as long as they do not become equal to zero.

If we analyze the propagation velocity of waves on water, it will be reveal that the velocity of waves depends on length of waves. The speed of long waves is proportional to the square root of the acceleration of gravity multiplied by the wave length:

$$v_\Phi = \sqrt{g\lambda}$$

The cause of these waves is the force of gravity.

For short waves the restoring force due to surface tension force, and therefore the speed of these waves is proportional to the square root of the private. The numerator of which is the surface tension, and in the denominator – the product of the wavelength to the density of water:

$$v_\Phi = \sqrt{\sigma / \lambda\rho}, \tag{64}$$

Suppose there is a channel with a constant slope bottom, extending to infinity along the axis OX.

And let the feed in a field of gravity flows, incompressible fluid. It is assumed that the fluid is devoid of internal friction. Friction neglects on the sides and bottom of the channel. The liquid level is above the bottom of the channel h. A small quantity compared with the characteristic dimensions of the flow, the size of the bottom roughness, etc.

Let $h = \xi + h_0$, \tag{65}

where h_0 ordinate denotes the free surface of the liquid (Fig. 6). Free liquid surface h_0 (Fig. 5), which is in equilibrium in the gravity field is flat. As a result of any external influence, liquid surface in a location removed from its equilibrium position. There is a movement spreading across the entire surface of the liquid in the form of waves, called gravity.

They are caused by the action of gravity field. This type of waves occurs mainly on the liquid surface. They capture the inner layers, the deeper for the smaller liquid surface.

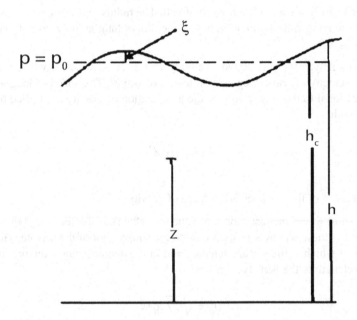

FIGURE 6 Fluid flow of with variable depth, where h_0 is the level of the free surface, ξ – A deviation from the level of the liquid free surface, h – Depth of the fluid and z – Vertical coordination of any point in the water column. We assume that the fluid flow is characterized by a spatial variable x and time dependent t.

Thus, it is believed that the fluid velocity u has a nonzero component u_x, which will be denoted by u (other components can be neglected). In addition, the level of h depends only on x and t.

Let us consider such gravitational waves, in which the speed of moving particles are so small that for the Euler equation, it can be neglected the $(u\nabla)u$ compared with $\partial u/\partial t$.

During the time period τ , committed by the fluid particles in the wave, these particles pass the distance of the order of the amplitude a .

Therefore, the speed of their movement will be $u \sim a/\tau$.

Rate u varies considerably over time intervals of the order τ and for distances of the order λ along the direction of wave propagation, λ Wavelength.

Therefore, the derivative of the velocity time – order u/τ and the coordinates – order u/λ .

Thus, the condition:

$$(u\nabla)u < \partial u/\partial t$$

Equivalent to the requirement

$$\frac{1}{\lambda}\left(\frac{a}{\lambda}\right)^2 < \frac{a}{\tau}\frac{1}{\tau} \quad a < \lambda, \qquad\qquad \text{or, (66)}$$

i.e., amplitude of the wave must be small compared with the wavelength.

Consider the propagation of waves in the channel OX directed along the axis for fluid flow along the channel.

Channel cross section can be of any shape and change along its length with changes in liquid level, cross-sectional area of the liquid in the channel denoted by: $h = h(x,t)$.

The depth of the channel and basin are assumed to be small compared with the wavelength.

We write the Euler equation in the form of

$$\frac{\partial u}{\partial t} = -\frac{1}{\rho}\frac{\partial p}{\partial x}, \qquad\qquad (67)$$

$$\frac{1}{\rho}\frac{\partial p}{\partial z} = -g,$$

(68)

where ρ – Density,

   p – Pressure,

   g – Acceleration of free fall.

Quadratic in velocity members omitted, since the amplitude of the waves is still considered low.

From the second equation we have that at the free surface:

$$z = h(x,t)$$

where, $p = p_0$ should be satisfied:

$$p = p_0 + \rho g(h - z),$$

(69)

Substituting this expression in Eq. (2), we obtain:

$$\frac{\partial u}{\partial t} = -g\frac{\partial h}{\partial x},$$

(70)

to determine u and h we use the continuity equation for the case under consideration.

Consider the volume of fluid contained between two planes of the cross-section of the canal at a distance dx from each other per unit time through a cross-section x enter the amount of fluid, equal to $(hu)_x$.

At the same time through the section:

$$x + dx$$

There is forth coming $(hu)_{x+dx}$.

Therefore, the volume of fluid between the planes is changed to

$$\left(hu\right)_{x+dx} - \left(hu\right)_x = \frac{\partial\left(hu\right)}{\partial x}dx \, , \qquad (71)$$

By virtue of incompressibility of the liquid is a change could occur only due to changes in its level. Changing the volume of fluid between these planes in a unit time is equal

$$\frac{\partial h}{\partial t}dx$$

Consequently, we can write:

$$\frac{\partial\left(hu\right)}{\partial x}dx = -\frac{\partial h}{\partial t}dx \text{ and } \frac{\partial\left(hu\right)}{\partial x} + \frac{\partial h}{\partial t} = 0, \, t > 0, \, -\infty < x < \infty \text{ or,} \qquad (72)$$

Since $h = h_0 + \xi$ where h_0 denotes the ordinate of the free liquid surface (Fig. 2), in a state of relative equilibrium and evolving the influence of gravity is:

$$\frac{\partial\xi}{\partial t} + h_0\frac{\partial u}{\partial x} = 0 \qquad (73)$$

Thus, we obtain the following system of equations describing the fluid flow in the channel:

$$\frac{\partial\xi}{\partial t} + h_0\frac{\partial u}{\partial x} = 0, \quad \frac{\partial u}{\partial t} + g\frac{\partial\xi}{\partial x} = 0, \, t > 0, \, -\infty < x < \infty, \qquad (74)$$

2.2.1 VELOCITY PHASE OF THE HARMONIC WAVE

The phase velocity h_0 expressed in terms of frequency v_Φ and wavelength f (or the angular frequency) λ and wave number $\omega = 2\pi f$ formula $k = 2\pi / \lambda$.

The concept of phase velocity can be used if the harmonic wave propagates without changing shape.

This condition is always performed in linear environments. When the phase velocity depends on the frequency, it is equivalent to talk about the velocity dispersion. In the absence of any dispersion the waves assumed with a rate equal to the phase velocity.

Experimentally, the phase velocity at a given frequency can be obtained by determining the wavelength of the interference experiments. The ratio of phase velocities in the two media can be found on the refraction of a plane wave at the plane boundary of these environments. This is because the refractive index is the ratio of phase velocities.

It is known that the wave number k satisfies the wave equation are not any values ω but only if their relationship. To establish this connection is sufficient to substitute the solution of the form:

$$\exp\left[i\left(\omega t - kx\right)\right],\tag{75}$$

in the wave equation.

The complex form is the most convenient and compact. We can show that any other representation of harmonic solutions, including in the form of a standing wave leads to the same connection between ω and k.

Substituting the wave solution into the equation for a string, we can see that the equation becomes an identity for:

$$\omega^2 = k^2 v_\Phi^2,\tag{76}$$

Exactly the same relation follows from the equations for waves in the gas, the equations for elastic waves in solids and the equation for electromagnetic waves in vacuum.

The presence of energy dissipation [Loytsyanskiy, L.G., Fluid, Moscow: Nauka, 1970, p.904] leads to the appearance of the first derivatives (forces of friction) in the wave equation. The relationship between frequency and wave number becomes the domain of complex numbers.

For example, the telegraph equation (for electric waves in a conductive line) yields:

$$\omega^2 = k^2 v_\phi^2 + i \cdot \omega R / L,$$ (77)

The relation connecting between a frequency and wave number (wave vector), in which the wave equation has a wave solution is called a dispersion relation, the dispersion equation or dispersion.

This type of dispersion relation determines the nature of the wave. Since the wave equations are partial differential equations of second order in time and coordinates, the dispersion is usually a quadratic equation in the frequency or wave number.

The simplest dispersion equations presented above for the canonical wave equation are also two very simple solutions:

$$\omega = +k v_\phi \text{ and } \omega = -k v_\phi,$$ (78)

We know that these two solutions represent two waves traveling in opposite directions. By its physical meaning the frequency is a positive value so that the two solutions must define two values of the wave number, which differ in sign. The Act permits the dispersion, generally speaking, the existence of waves with all wave numbers that is of any length, and, consequently, any frequencies. The phase velocity of these waves:

$$v_\Phi = \omega / k,$$ (79)

Coincides with the most velocity, which appears in the wave equation and is a constant that depends only on the properties of the medium.

The phase velocity depends on the wave number, and, consequently, on the frequency. The dispersion equation for the telegraph equation is an algebraic quadratic equation has complex roots. By analogy with the theory of oscillations, the presence of imaginary part of the frequency means the damping or growth of waves. It can be noted that the form of the dispersion law determines the presence of damping or growth.

In general terms, the dispersion can be represented by the equation: $\Phi(\omega, \bar{k}) = 0$ where Φ – A function of frequency and wave vector.

By solving this equation for ω you can obtain an expression for the phase velocity :

$$v_\Phi = \omega / k = f\left(\omega, \vec{k}\right),$$

(80)

By definition, the phase velocity is a vector directed normal to phase surface. Then, more correctly write the last expression in the following form:

$$\vec{v}_\Phi = \frac{\lambda}{T} = \frac{\omega}{k^2} \cdot \vec{k} = f\left(\omega, \vec{k}\right),$$

(81)

2.2.2 DISPERSIVE PROPERTIES OF MEDIA

The most important subject of research in wave physics, which has the primary practical significance.

If we refer to dimensionless parameters and variables:

$$\tau = t\sqrt{\frac{g}{h_0}}, X = \frac{x}{h_0}, U = u\frac{1}{\sqrt{gh_0}}, \delta = \frac{\xi}{h_0},$$

(82)

The system of Eq. (8) becomes:

$$\frac{\partial \delta}{\partial \tau} + \frac{\partial U}{\partial X} = 0, \frac{\partial U}{\partial \tau} + \frac{\partial \delta}{\partial X} = 0, t > 0, -\infty < X < \infty,$$

(83)

Consider a plane harmonic longitudinal waves, i.e., we seek the solution of Eq. (9) as the real part of the following complex expressions:

$$\Psi = \Psi^0 \exp\left[i\left(k_*X + \omega_* \tau\right)\right], \quad \Psi^0 = \Psi^0_* + i\Psi^0_{**}, \left|\Psi^0\right| \ll 1$$

$$k_* = k + ik_{**}, \omega_* = \omega + i\omega_{**},$$

(84)

where,

$$\Psi = \delta, U \text{ , a } \Psi^0 = \delta^0, U^0$$

determines the amplitude of the perturbations of displacement and velocity.

There are two types of solutions.

Type I. Solution or wave of the first type, when:

$k_* = k$ – A real positive number $(k > 0, k_{**} = 0)$..

In this case we have:

$$\Psi = \left(\Psi_*^0 + i\Psi_{**}^0\right)\exp\left[i\left(kX + \omega\tau + i\omega_{**}\tau\right)\right] = \left(\Psi_*^0 + i\Psi_{**}^0\right)\exp\left(-\omega_{**}\tau\right) \times$$

$$\left[\cos\left(kX + \omega\tau\right) + i\sin\left(kX + \omega\tau\right)\right], \tag{85}$$

$$\mathrm{Re}\{\Psi\} = \exp\left(-\omega_{**}\tau\right)\left|\Psi^0\right|\sin\left[\phi + \left(kX + \omega\tau\right)\right], \tag{86}$$

$$\left|\Psi^0\right| = \sqrt{\Psi_*^{0^2} + \Psi_{**}^{0^2}} \text{ , } \varphi = arctg\left(-\Psi_*^0 / \Psi_{**}^0\right)$$

Thus, the decision of the first type is a sinusoidal coordinate and $\omega_{**} > 0$ decaying exponentially in time perturbation, which is called k – wave:

$$\Psi(k) = \left|\Psi^0\right|\exp\left[-\omega_{**}(k)\tau\right]\sin\left\{\phi + \frac{2\pi\left[X + v_\Phi(k)\tau\right]}{\lambda(k)}\right\}, \tag{87}$$

where,

$$v_\Phi(k) = \omega(k)/k, \ \lambda(k) = 2\pi/k, \tag{88}$$

φ – Initial phase

Here,

$v_{\Phi}(k)$ – phase velocity or the velocity of phase fluctuations,

$\lambda(k)$ – Wavelength,

$\omega_{**}(k)$ – damping the oscillations in time.

In other words,

k – Waves have uniform length, but time-varying amplitude.

These waves are analog of free oscillations.

Type II. Decisions, or wave, the second type, when:

$\omega_* = \omega$ – a

Real positive number $(\omega > 0, \omega_{**} = 0)$.

In this case we have:

$$\psi = \left(\psi_*^0 + i\psi_{**}^0\right)\exp\left[i\left(kX + \omega\tau + ik_{**}z\right)\right] = \left(\Psi_*^0 + i\Psi_{**}^0\right)\exp\left(-k_{**}X\right)\times$$

$$\left[\cos\left(kX + \omega\tau\right) + i\sin\left(kX + \omega\tau\right)\right], \tag{89}$$

$$\operatorname{Re}\{\Psi\} = \exp\left(-k_{**}X\right)\left|\Psi^0\right|\sin\left[\phi + \left(kX + \omega\tau\right)\right], \tag{90}$$

Thus, the solution of the second type is a sinusoidal oscillation in time (excited, for example, any stationary source of external monochromatic vibrations at) $X = 0$, decaying exponentially along the length of the amplitude.

Such disturbances, which are analogous to a wave of forced oscillations, called ω – waves:

$$\Psi(\omega) = \left|\Psi^0(\omega)\right|\exp\left(-k_{**}(\omega)X\right)\sin\left\{\phi + \frac{2\pi\left[X + v_{\Phi}(\omega)\tau\right]}{\lambda(\omega)}\right\}, \tag{91}$$

$$v_\Phi\left(w\right) = w \,/\, k\left(w\right),$$ (92)

$$\lambda(\omega) = 2\pi \,/\, k(\omega)$$

Here, $k_{**}\left(\omega\right)$ – damping vibrations in length.

In other words, ω – waves – waves with stationary in time but varying in length amplitudes.

Cases $k < 0, k_{**} > 0$ and $k > 0, k_{**} < 0$ consistent with attenuation of amplitude of the disturbance regime in the direction of phase fluctuations or phase velocity.

Let us obtain the characteristic equation, linking k_* and ω_*.

After substituting Eq. (10) in the system of Eq. (9) we obtain:

$$\delta^0 \frac{\omega_*}{k_*} + U^0 = 0, \quad U^0 \frac{\omega_*}{k_*} + \delta^0 = 0,$$ (93)

From the condition of the existence of a system of linear homogeneous algebraic Eq. (13) with respect to perturbations of a nontrivial solution implies the desired characteristic, or dispersion, which has one solution:

$$v_\Phi = \sqrt{gh_0},$$ (94)

Thus, we obtain a solution representing a sinusoidal in time and coordinate free undammed oscillations.

Such behaviors of the waves are due to the absence of any dissipation in the fluid. The fluid is incompressible and ideal. There is no heat – mass transfer.

Eq. (9) with respect to perturbations take the form of wave equations:

$$\frac{\partial^2 \xi}{\partial t^2} = gh_0 \frac{\partial^2 \xi}{\partial x^2} \text{ and } \frac{\partial^2 u}{\partial t^2} = gh_0 \frac{\partial^2 u}{\partial x^2},$$ (95)

Note that in gas dynamics $v_\Phi = \sqrt{gh_0}$ equivalent to the speed of sound.

2.3   CONCLUSION

Thus, we obtain a solution representing a sinusoidal in time and coordinate free undammed oscillations. Such behaviors of the waves are due to the absence of any dissipation in the fluid. The fluid is incompressible and ideal. There is no heat mass transfer.

KEYWORDS

- **Dispersed fluid**
- **Hydraulic system**
- **Ideal fluid**
- **Incompressible fluid**
- **Pipe wall**

REFERENCES

1. Hariri Asli, K.; GIS Water hammer disaster at earthquake in Rasht water pipeline, 3rd International Conference on Integrated Natural Disaster Management, INDM2008, http://www.civilica.com/Paper-INDM03-INDM03_001.html.
2. Hariri Asli, K.; Nagiyev, F. B.; Beglou, M. J.; Haghi, A. K.; Kinetic analysis of convective drying, International Journal of the Balkan Tribological Association, ISSN: 1310–4772, Sofia, Bulgaria, **2009**, *15(4),* 546–556, jbalkta@gmail.com
3. HaririAsli, K.; Nagiyev, F. B.; Bubbles characteristics and convective effects in the binary mixtures. Transactions issue mathematics and mechanics series of physical-technical and mathematics science, ISSN 0002–3108, Azerbaijan, Baku, 215–220, **2008,** www.imm.science.az/journals/AMEA_xeberleri/.../215–220.pdf
4. Hariri Asli, K.; Nagiyev, F. B.; Haghi, A. K.; Three-dimensional Conjugate Heat Transfer in Porous Media, International Journal of the Balkan Tribological Association, ISSN: 1310–4772, Sofia, Bulgaria, **2009,** *15(3),* 336–346, jbalkta@gmail.com
5. Hariri Asli, K.; Nagiyev, F. B.; Haghi, A. K.; Water hammer and fluid condition; a computational approach, Computational Methods in Applied Science and Engineering, USA, Chapter *5,* Nova Science Publications, ISBN: 978-1-60876-052-7, USA, **2010,** 73–94, https://www.novapublishers.com/catalog/
6. Hariri Asli, K.; Nagiyev, F. B.; Haghi, A. K.; Interpenetration of two fluids at parallel between plates and turbulent moving in pipe; a case study, Computational Methods in Applied Science and Engineering, USA, Chapter 7, Nova Science Publications,

ISBN: 978-1-60876-052-7, USA, **2010**, 107–133, https://www.novapublishers.com/catalog/

7. Hariri Asli, K.; Nagiyev, F. B.; Haghi, A. K.; Modeling for water hammer due to valves; from theory to practice, Computational Methods in Applied Science and Engineering, USA, Chapter 11, Nova Science Publications ISBN: 978-1-60876-052-7, USA, **2010**, 229–236, https://www.novapublishers.com/catalog/

8. Hariri Asli, K.; Nagiyev, F. B.; Haghi, A. K.; A computational method to Study transient flow in binary mixtures, Computational Methods in Applied Science and Engineering, USA, Chapter 13, Nova Science Publications ISBN: 978-1-60876-052-7, USA, **2010**, 229–236, https://www.novapublishers.com/catalog/

9. Hariri Asli, K.; Nagiyev, F. B.; Haghi, A. K.; Water hammer analysis; some computational aspects and practical hints, Computational Methods in Applied Science and Engineering, USA, Chapter 16, Nova Science Publications ISBN: 978-1-60876-052-7, USA, **2010**, 263–281, https://www.novapublishers.com/catalog/

10. Hariri Asli, K.; Nagiyev, F. B.; Haghi, A. K.; Water hammer and hydrodynamics instabilities modeling, Computational Methods in Applied Science and Engineering, USA, Chapter 17, From Theory to Practice, Nova Science Publications ISBN: 978-1-60876-052-7, USA, **2010**, 283–301, https://www.novapublishers.com/catalog/

11. Hariri Asli, K.; Nagiyev, F. B.; Haghi, A. K.; A computational approach to study water hammer and pump pulsation phenomena, Computational Methods in Applied Science and Engineering, USA, Chapter 22, Nova Science Publications, ISBN: 978-1-60876-052-7, USA, **2010**, 349–363, https://www.novapublishers.com/catalog/

12. Hariri Asli, K.; Nagiyev, F. B.; Haghi, A. K.; Some aspects of physical and numerical modeling of water hammer in pipelines. Computational Methods in Applied Science and Engineering, USA, Chapter 23, Nova Science Publications, ISBN: 978-1-60876-052-7, USA, **2010**, 365–387, https://www.novapublishers.com/catalog/

13. Hariri Asli, K.; Nagiyev, F. B.; Haghi, A. K.; A computational approach to study fluid movement, Nanomaterials Yearbook – **2009,** From Nanostructures, Nanomaterials and Nanotechnologies to Nanoindustry, Chapter 16, Nova Science Publications, USA, ISBN: 978-1-60876-451-8, USA, **2010**, 181–196. https://www.novapublishers.com/catalog/product_info.php?products_id=11587

14. Hariri Asli, K.; Nagiyev, F. B.; Haghi, A. K.; Physical modeling of fluid movement in pipelines, Nanomaterials Yearbook –2009, From Nanostructures, Nanomaterials and Nanotechnologies to Nanoindustry, Chapter 17, Nova Science Publications, USA, ISBN: 978-1-60876-451-8, USA, **2010**, 197–214, https://www.novapublishers.com/catalog/product_info.php? products_id=11587

15. Hariri Asli, K.; Nagiyev, F. B.; Haghi, A. K.; Aliyev, S. A.; Improved modeling for prediction of water transmission failure, Recent Progress in Research in Chemistry and Chemical Engineering, Chapter 2, Nova Science Publications, ISBN: 978-1-61668-501-0, Nova Science Publications, USA, **2010, 28–36**. https://www.novapublishers.com/catalog/product_info.php?products_id=13174

16. Hariri Asli, K.; Nagiyev, F. B.; Haghi, A. K.; Aliyev S. A.; Pure Oxygen penetration in wastewater flow, Recent Progress in Research in Chemistry and Chemical Engineering, Chapter 3, Nova Science Publications, ISBN: 978-1-61668-501-0, Nova Science Publications, USA, 17–27, **2010**, https://www.novapublishers.com/catalog/product_info.php?products_id=13174

17. Hariri Asli, K.; Mathematics and numerical modeling Technology, Journal of Mathematics and Technology, ISSN: 2078–0257, No.3, August, Baku, Azerbaijan, 68–74, **2010,** https://www.International%20Journal%20of%20Academic%20Research-IJAR. htm

18. Hariri Asli, K.; Nagiyev, F. B.; Haghi, A. K.; Aliyev, S. A.; Physical and Numerical Modeling of Fluid Flow in Pipelines: A computational approach, International Journal of the Balkan Tribological Association, ISSN: 1310–4772, *16(1),* Sofia, Bulgaria, 20–34, **2010,** jbalkta@gmail.com

19. Hariri Asli, K.; Nagiyev, F. B.; Haghi, A. K.; Aliyev, S. A.; A Numerical Study on heat transfer in Microtubes, International Journal of the Balkan Tribological Association, ISSN: 1310–4772, 16, *1,* Sofia, Bulgaria, 9–19, **2010,** jbalkta@gmail.com

20. Hariri Asli, K.; Nagiyev, F. B.; Haghi, A. K.; A numerical study on fluid dynamics, Material Science Synthesis, Properties, Applicators, ISBN: 978-1-60876-872-1, Chapter 15, Nova Science Publications, USA, **2010,** 101–110, https://www.novapublishers.com/catalog/product_info.php?products_id=12129

21. Hariri Asli, K.; Nagiyev, F. B.; Haghi, A. K.; Some interpenetration for turbulent moving of fluid in pipe, Material Science Synthesis, Properties, Applicators, ISBN: 978-1-60876-872-1, Chapter 16, Nova Science Publications, USA, **2010,** 111–117. https://www.novapublishers.com/catalog/product_info.php?products_id=12129

22. Hariri Asli, K.; Nagiyev, F. B.; Haghi, A. K.; Fluid flow analysis due to water hammer, Material Science Synthesis, Properties, Applicators, ISBN: 978-1-60876-872-1, Chapter 17, Nova Science Publications, USA, **2010,** 120–128. https://www.novapublishers.com/catalog/product_info.php?products_id=12129

23. Hariri Asli, K.; Nagiyev, F. B.; Haghi, A. K.; Transient flow in binary mixtures, Material Science Synthesis, Properties, Applicators, ISBN: 978-1-60876-872-1, Chapter 19, Nova Science Publications, USA, **2010,** 164–176. https://www.novapublishers.com/catalog/product_info.php?products_id=12129

24. Hariri Asli, K.; Nagiyev, F. B.; Haghi, A. K.; Hydrodynamics instabilities modeling, Material Science Synthesis, Properties, Applicators, ISBN: 978-1-60876-872-1, Chapter 20 Nova Science Publications, USA, **2010,** 140–146. https://www.novapublishers.com/catalog/product_info.php?products_id=12129

25. Hariri Asli, K.; Nagiyev, F. B.; Haghi, A. K.; Fluid dynamics and pump pulsation, Material Science Synthesis, Properties, Applicators, ISBN: 978-1-60876-872-1, Chapter 21, Nova Science Publications, USA, **2010,** 147–155. https://www.novapublishers.com/catalog/product_info.php?products_id=12129

26. Hariri Asli, K.; Nagiyev, F. B.; Haghi, A. K.; Aliyev, S. A.; Hariri Asli H.; Flow in water pipeline: A computational approach, International Journal of Academic Research, ISSN: 1310–4772, ISSN: 2075–4124, *2(5),* September 30, Baku, Azerbaijan, 164–176, **2010,** https://www.International%20Journal%20of%20Academic%20Research-IJAR.htm

27. Hariri Asli, K.; Nagiyev, F. B.; Haghi, A. K.; Aliyev, S. A.; Nonlinear Heterogeneous Model for Water Hammer Disaster, International Journal of the Balkan Tribological Association, ISSN: 1310–4772, *16(2),* Sofia, Bulgaria, 209–222, **2010,** jbalkta@gmail.com

28. Hariri Asli, K.; Nagiyev, F. B.; Haghi, A. K.; Heat flow and mass transfer in capillary Porous body, Journal of the Balkan Tribological Association, *16(3)*, Tribotechnics and tribomechanics, Sofia, Bulgaria, 353–361, **2010**, jbalkta@gmail.com

29. Hariri Asli, K.; Nagiyev, F. B.; Haghi, A. K.; A Numerical Study on thermal drying of Porous solid, Journal of the Balkan Tribological Association, *16(3)*, Tribotechnics– thermal drying, Sofia, Bulgaria, 373–381, 2010, jbalkta@gmail.com

30. Hariri Asli, K.; Haghi, A. K.; A Numerical Study on Fluid Flow and Pressure drop in Microtubes, Journal of the Balkan Tribological Association, *16(3)*, Tribotechnics and tribomechanics, Sofia, Bulgaria, 382–392, **2010**, jbalkta@gmail.com

31. Hariri Asli, K.; Nagiyev, F. B.; Haghi, A. K.; Aliyev, S. A.; Hariri Asli, H.; Improved Nonlinear Heterogeneous Model for Wastewater Treatment, International Journal on "Technical and Physical Problems of Engineering," (IJTPE), Published by the International Organization on TPE (IOTPE), ISSN: 2077–3528, Baku, Azerbaijan, **2010**, 30–36, http://www.iotpe.com/TPE-Journal/PublicationPolicy.html

32. Hariri Asli, K.; GIS Nonlinear Dynamics Model: Some Computational Aspects and Practical Hints, International Journal on "Technical and Physical Problems of Engineering," (IJTPE), Published by the International Organization on TPE (IOTPE), ISSN: 2077–3528, 1–5, Baku, Azerbaijan, **2010**, http://www.iotpe.com/TPE-Journal/ PublicationPolicy.html

33. Skousen P.; "Valve Handbook," McGraw Hill, New York, HAMMER Theory and Practice, **1998**, 687–721.

34. Shaking, N. I.; Water hammer to break the continuity of the flow in pressure conduits pumping stations: Dis. on Kharkov, **1988**, 225.

35. Tijsseling,"Alan E Vardy Time scales and FSI in unsteady liquid-filled pipe flow," **1993**, 5–12.

36. Wu, P. Y.; Little, W. A.; Measurement of friction factor for flow of gases in very fine channels used for micro miniature, Joule Thompson refrigerators, Cryogenics *24(8)*, **1983**, 273–277.

37. Song C. C. et al.; "Transient Mixed-Flow Models for Storm Sewers,"*J. Hyd. Div.* Nov.; **1983**, *109*, 458–530.

38. Stephenson D.; "Pipe Flow Analysis," Elsevier, *19*, S.A. **1984**, 670–788.

39. Chaudhry, M. H.; "Applied Hydraulic Transients," Van Nostrand Reinhold Co.: N.Y. **1979**, 1322–1324.

40. Chaudhry, M. H.; Yevjevich V. "Closed Conduit Flow," Water Resources Publication, USA, **1981**, 255–278.

41. Chaudhry, M. H.; Applied Hydraulic Transients, Van Nostrand Reinhold, New York, USA, **1987**, 165–167.

42. Kerr, S. L.; "Minimizing service interruptions due to transmission line failures: Discussion," Journal of the American Water Works Association, *41, 634,* July **1949**, 266–268.

43. Kerr, S. L.; "Water hammer control," Journal of the American Water Works Association, *43,* December **1951**, 985–999.

44. Apoloniusz Kodura, Katarzyna Weinerowska,"Some Aspects of Physical and Numerical Modeling of Water Hammer in Pipelines," **2005**, 126–132.

45. Anuchina, N. N.; Volkov V. I.; Gordeychuk V. A.; Es'kov, N. S.; Ilyutina, O. S.; Kozyrev O. M. "Numerical simulations of Rayleigh-Taylor and Richtmyer-Meshkov instability using mah-3 code,"*J.Comput. Appl. Math.***2004,** *168,* 11.

46. Fox, J. A.; "Hydraulic Analysis of Unsteady Flow in Pipe Network," Wiley: N.Y. **1977,** 78–89.

47. Karassik, I. J.; "Pump Handbook – Third Edition," McGraw-Hill, **2001,** 19–22.

48. Fok, A.; "Design Charts for Air Chamber on Pump Pipelines,"*J. Hyd. Div.ASCE,* Sept.; **1978,** 15–74.

49. Fok, A.; Ashamalla A.; Aldworth G.; "Considerations in Optimizing Air Chamber for Pumping Plants," Symposium on Fluid Transients and Acoustics in the Power Industry, San Francisco, USA, Dec, **1978,** 112–114.

50. Fok, A.; "Design Charts for Surge Tanks on Pump Discharge Lines," BHRA 3rd Int. Conference on Pressure Surges, Bedford, England, Mar.; **1980,** 23–34.

51. Fok, A.; "Water hammer and Its Protection in Pumping Systems," Hydro technical Conference, CSCE, Edmonton, May,**1982,** 45–55.

52. Fok, A.; "A contribution to the Analysis of Energy Losses in Transient Pipe Flow," PhD Thesis, University of Ottawa, **1987,** 176–182.

53. Hariri Asli, K.; Nagiyev, F. B.; Water Hammer and fluid condition, Ministry of Energy, Gilan Water and Wastewater Co.; Research Week Exhibition, Tehran, Iran, December, **2007,** 132–148, http://isrc.nww.co.ir.

54. Hariri Asli, K.; Nagiyev, F. B.; Water Hammer analysis and formulation, Ministry of Energy, Gilan Water and Wastewater Co.; Research Week Exhibition, Tehran, Iran, December, **2007,** 111–131, http://isrc.nww.co.ir.

55. Hariri Asli, K.; Nagiyev, F. B.; Water Hammer and hydrodynamics instabilities, Interpenetration of two fluids at parallel between plates and turbulent moving in pipe, Ministry of Energy, Guilan Water and Wastewater Co.; Research Week Exhibition, Tehran, Iran, December, **2007,** 90–110, http://isrc.nww.co.ir.

56. Hariri Asli, K.; Nagiyev, F. B.; Water Hammer and pump pulsation, Ministry of Energy, Guilan Water and Wastewater Co.; Research Week Exhibition, Tehran, Iran, December, **2007,** 51–72, http://isrc.nww.co.ir.

57. Hariri Asli, K.; Nagiyev, F. B.; Reynolds number and hydrodynamics' instability," Ministry of Energy, Guilan Water and Wastewater Co.; Research Week Exhibition, Tehran, Iran, December, **2007,** 31–50, http://isrc.nww.co.ir.

58. Hariri Asli, K.; Nagiyev, F. B.; Water Hammer and valves, Ministry of Energy, Guilan Water and Wastewater Co.; Research Week Exhibition, Tehran, Iran, December, **2007,** 20–30, http://isrc.nww.co.ir.

59. Hariri Asli, K.; Nagiyev, F. B.; "Interpenetration of two fluids at parallel between plates and turbulent moving in pipe," Ministry of Energy, Guilan Water and Wastewater Co.; Research Week Exhibition, Tehran, Iran, December, **2007,** 73–89, http:// isrc.nww.co.ir.

60. Hariri Asli, K.; Nagiyev, F. B.; Decreasing of Unaccounted For Water "UFW" by Geographic Information System"GIS" in Rasht urban water system, civil engineering organization of Guilan, Technical and Art Journal, **2007,** 3–7, http://www.art-of-music.net/.

61. Hariri Asli, K.; Portable Flow meter Tester Machine Apparatus, Certificate on registration of invention, Tehran, Iran, #010757, Series a/82, 24/11/2007, 1–3.

62. Hariri Asli, K.; Nagiyev, F. B.; Haghi, A. K.; "Interpenetration of two fluids at parallel between plates and turbulent moving in pipe," 9th Conference on Ministry of Energetic works at research week, Tehran, Iran, **2008**, 73–89, http://isrc.nww.co.ir.

63. Hariri Asli, K.; Nagiyev, F. B.; Haghi, A. K.; "Water hammer and valves," 9th Conference on Ministry of Energetic works at research week, Tehran, Iran, **2008**, 20–30, http://isrc.nww.co.ir.

64. Hariri Asli, K.; Nagiyev, F. B.; Haghi, A. K.; "Water hammer and hydrodynamics instability," 9th Conference on Ministry of Energetic works at research week, Tehran, Iran, **2008**, p.90–110, http://isrc.nww.co.ir.

65. Hariri Asli, K.; Nagiyev, F. B.; Haghi, A. K.; "Water hammer analysis and formulation," 9th Conference on Ministry of Energetic works at research week, Tehran, Iran, **2008**, p. 27–42, http://isrc.nww.co.ir.

66. Hariri Asli, K.; Nagiyev, F. B.; Haghi, A. K.; "Water hammer & fluid condition," 9th Conference on Ministry of Energetic works at research week, Tehran, Iran, **2008**, 27–43, http://isrc.nww.co.ir.

67. Hariri Asli, K.; Nagiyev, F. B.; Haghi, A. K.; "Water hammer and pump pulsation," 9th Conference on Ministry of Energetic works at research week, Tehran, Iran, **2008**, 27–44, http://isrc.nww.co.ir.

68. Hariri Asli, K.; Nagiyev, F. B.; Haghi, A. K.; "Reynolds number and hydrodynamics instability," 9th Conference on Ministry of Energetic works at research week, Tehran, Iran, **2008**, 27–45, http://isrc.nww.co.ir.

69. Hariri Asli, K.; Nagiyev, F. B.; Haghi, A. K.; "Water hammer and fluid Interpenetration," 9th Conference on Ministry of Energetic works at research week, Tehran, Iran, **2008**, 27–47, http://isrc.nww.co.ir.

70. Hariri Asli, K.; GIS and water hammer disaster at earthquake in Rasht water pipeline, civil engineering organization of Guilan, Technical and Art Journal, **2008**, 14–17, http://www.art-of-music.net/.

71. Hariri Asli, K.; GIS and water hammer disaster at earthquake in Rasht water pipeline, 3rd International Conference on Integrated Natural Disaster Management, Tehran university, ISSN: 1735–5540, 18–19 Feb.; INDM, Tehran, Iran, **2008**, *13*, 53/1–12, http://www.civilica.com/Paper-INDM03-INDM03_001.html

72. Hariri Asli, K.; Nagiyev, F. B.; Bubbles characteristics and convective effects in the binary mixtures. Transactions issue mathematics and mechanics series of physical-technical and mathematics science, ISSN: 0002–3108, Azerbaijan, Baku, **2009**, 68–74, http://www.imm.science.az/journals.html.

73. Hariri Asli, K.; Nagiyev, F. B.; Haghi, A. K.; Aliyev, S. A.; Three-Dimensional conjugate heat transfer in porous media, 1st Festival on Water and Wastewater Research and Technology, Tehran, Iran, 12–17 Dec. **2009**, 26–28, http://isrc.nww.co.ir.

74. Hariri Asli, K.; Nagiyev, F. B.; Haghi, A. K.; Aliyev, S. A.; Some Aspects of Physical and Numerical Modeling of water hammer in pipelines, 1st Festival on Water and Wastewater Research and Technology, Tehran, Iran, 12–17 Dec. **2009**, 26–29, http://isrc.nww.co.ir

75. Hariri Asli, K.; Nagiyev, F. B.; Haghi, A. K.; Aliyev, S. A.; Modeling for Water Hammer due to valves: From theory to practice, 1st Festival on Water and Wastewater Research and Technology, Tehran, Iran, 12–17 Dec. **2009**, 26–30, http://isrc.nww.co.ir.

76. Hariri Asli, K.; Nagiyev, F. B.; Haghi, A. K.; Aliyev, S. A.; Water hammer and hydro-dynamics instabilities modeling: From Theory to Practice, 1st Festival on Water and Wastewater Research and Technology, Tehran, Iran, 12–17 Dec. **2009**, 26–31, http://isrc.nww.co.ir

77. Hariri Asli, K.; Nagiyev, F. B.; Haghi, A. K.; Aliyev, S. A.; A computational approach to study fluid movement, 1st Festival on Water and Wastewater Research and Technology, Tehran, Iran, 12–17 Dec. **2009**, 27–32, http://isrc.nww.co.ir.

78. Hariri Asli, K.; Nagiyev, F. B.; Haghi, A. K.; Aliyev, S. A.; Water Hammer Analysis: Some Computational Aspects and practical hints, 1st Festival on Water and Wastewater Research and Technology, Tehran, Iran, 12–17 Dec. **2009**, 27–33, http://isrc.nww.co.ir

79. Hariri Asli, K.; Nagiyev, F. B.; Haghi, A. K.; Aliyev, S. A.; Water Hammer and Fluid condition: A computational approach, 1st Festival on Water and Wastewater Research and Technology, Tehran, Iran, 12–17 Dec.; 2009, 27–34, http://isrc.nww.co.ir.

80. Hariri Asli, K.; Nagiyev, F. B.; Haghi, A. K.; Aliyev, S. A.; A computational Method to Study Transient Flow in Binary Mixtures, 1st Festival on Water and Wastewater Research and Technology, Tehran, Iran, 12–17 Dec. **2009**, 27–35, http://isrc.nww.co.ir.

81. Hariri Asli, K.; Nagiyev, F. B.; Haghi, A. K.; Physical modeling of fluid movement in pipelines, 1st Festival on Water and Wastewater Research and Technology, Tehran, Iran, 12–17 Dec. **2009**, 27–36, http://isrc.nww.co.ir.

82. Hariri Asli, K.; Nagiyev, F. B.; Haghi, A. K.; Aliyev, S. A.; Interpenetration of two fluids at parallel between plates and turbulent moving, 1st Festival on Water and Wastewater Research and Technology, Tehran, Iran, 12–17 Dec. **2009**, 27–37, http://isrc.nww.co.ir.

83. Hariri Asli, K.; Nagiyev, F. B.; Haghi, A. K.; Aliyev, S. A.; Modeling of fluid interaction produced by water hammer, 1st Festival on Water and Wastewater Research and Technology, Tehran, Iran, 12–17 Dec. **2009**, 27–38, http://isrc.nww.co.ir.

84. Hariri Asli, K.; Nagiyev, F. B.; Haghi, A. K.; Aliyev, S. A.; GIS and water hammer disaster at earthquake in Rasht pipeline, 1st Festival on Water and Wastewater Research and Technology, Tehran, Iran, 12–17 Dec. **2009**, 27–39, http://isrc.nww.co.ir.

85. Hariri Asli, K.; Nagiyev, F. B.; Haghi, A. K.; Aliyev, S. A.; Interpenetration of two fluids at parallel between plates and turbulent moving, 1st Festival on Water and Wastewater Research and Technology, Tehran, Iran, 12–17 Dec. **2009**, 27–40, http://isrc.nww.co.ir.

86. Hariri Asli, K.; Nagiyev, F. B.; Haghi, A. K.; Aliyev, S. A.; Water hammer and hydro-dynamics' instability, 1st Festival on Water and Wastewater Research and Technology, Tehran, Iran, 12–17 Dec. **2009**, 27–41, http://isrc.nww.co.ir.

87. Hariri Asli, K.; Nagiyev, F. B.; Haghi, A. K.; Aliyev, S. A.; Water hammer analysis and formulation, 1st Festival on Water and Wastewater Research and Technology, Tehran, Iran, 12–17 Dec. **2009**, 27–42, http://isrc.nww.co.ir.

88. Hariri Asli, K.; Nagiyev, F. B.; Haghi, A. K.; Aliyev, S. A.; Water hammer &fluid condition, 1st Festival on Water and Wastewater Research and Technology, Tehran, Iran, 12–17 Dec. 2009, 27–43, http://isrc.nww.co.ir.

89. Hariri Asli, K.; Nagiyev, F. B.; Haghi, A. K.; Aliyev, S. A.; Water hammer and pump pulsation, 1st Festival on Water and Wastewater Research and Technology, Tehran, Iran, 12–17 Dec. 2009, 27–44, http://isrc.nww.co.ir.
90. Hariri Asli, K.; Nagiyev, F. B.; Haghi, A. K.; Aliyev, S. A.; Reynolds number and hydrodynamics instabilities, 1st Festival on Water and Wastewater Research and Technology, Tehran, Iran, 12–17 Dec. 2009, 27–45, http://isrc.nww.co.ir.
91. Hariri Asli, K.; Nagiyev, F. B.; Haghi, A. K.; Aliyev, S. A.; water hammer and valves, 1st Festival on Water and Wastewater Research and Technology, Tehran, Iran, 12–17 Dec. 2009, 27–46, http://isrc.nww.co.ir.
92. Hariri Asli, K.; Nagiyev, F. B.; Haghi, A. K.; Aliyev, S. A.; "Water hammer and fluid Interpenetration," 1st Festival on Water and Wastewater Research and Technology, Tehran, Iran, 12–17 Dec. 2009, 27–47, http://isrc.nww.co.ir.

CHAPTER 3

MODELING FOR PRESSURE WAVE INTO WATER PIPELINE

CONTENTS

3.1 INTRODUCTION

The development of hydraulic transition was paid more attention by N. E. Zhukovsky, A. Surin, L. Bergeron, L.F. Moshnin, N. A. Kartvelishvili, M. Andriashev, V. S. Dikarevsky, K.P. Wisniewski, B. F. Laman, V.I. Blokhin, L. S. Gerashchenko, V. N. Kovalenko, and others. The most detailed experimental and theoretical study of water hammer with a discontinuity in the flow conduits performed by D. N. Smirnov and L. B. Zubov. As a result of the research, they describe the basic laws of gap columns, fluid and obtained relatively simple calculation dependences. In the above works, there are methods of determining maximal pressures after the discontinuities of the flow. However, the results of calculations by these methods are often contradictory. In addition, not clarified the conditions under which the maximum pressure generated. There is little influence of loss of pressure, vacuum, nature and duration of flow control and other factors on the value of maximum pressure. The study of V. S. Dikarevsky for water hammer was included to break the continuity of flow. His work dealt with in detail, the impact magnitude of the vacuum on the course of the entire process of water hammer. Analytically and based on experimental data, scholars have argued that in a horizontal pipe rupture. The continuity of the flow occurs mainly in the regulatory body, and cavitation phenomena on the length of the pipeline are manifested. It investigates only in the form of small bubbles, whose influence on the process of hydraulic impact is negligible. As a result, research scientists have obtained analytic expressions for the hydraulic shock. They mention a gap of continuous flow, taking into account the energy loss, while controlling the flow and the wave nature. However, studies of V. S. Dikarevskogo were conducted mainly for the horizontal pressure pipelines and pumping units with a low inertia of moving masses. Researches of N. I. Kolotilo and others devoted to the study of water hammer to break the continuity of flow in the intermediate point. N. I. Kolotilo analytically derived the condition for the gap of continuous flow at a turning point of the pipeline when the pressure is reduced at this point (below atmospheric pressure). Studies have shown that the location of the discontinuity of continuous flow at a turning point depends, first of all, the profile of the pipeline. Protection of hydraulic systems against water hammer by releasing part of the transported fluid is the most widespread method of artificial reduction of the hydraulic shock. Devices that perform this function can be divided into valve, bursting disc and the overflow of the column. Development of algorithms for software simulation of transients by K. P. Vishnevsky was made for the complex pressure systems. It included the possible formation of discontinuities flows, hydraulic resistance, structural features of the pumping of water systems (pumps, piping, valves, etc.). However, a calculation of water ham

mer is adapted to high-pressure water systems for household and drinking purposes. K. P. Vishnevsky used "characteristics method" for the calculation of
water hammer on a computer dedicated to the work of B. F. Lyamaeva [1-35].

3.2   MATERIAL AND METHODS

The velocity $V(z, t)$ becomes dependent on the axial position z, and a continuity equation must be considered in addition to the equation of motion.
Present research assumed, water hammer pressure or surge pressure (H) is
a function of independent variables (X) such as: $H \approx$, Cl, Ep, Ew, V, T, C,
g, Tp, f, g, D, L.

Transient events data on water transmission line have been collected first: at
starting-point of water hammer condition and second: at water hammer condition. The main approach in this research has focused oninvestigation of relation
between: P-surge pressure (m), as a function"Y," and several factors, as variables
"X," such as: -density (kg/m³), C-velocity of surge wave (m/s), g-acceleration of
gravity (ms⁻²), V-changes in velocity of water (m/s), d-pipe diameter (m), T-pipe
thickness (m), Ep-pipe module of elasticity (kg/m²), Ew-module of elasticity of
water(kg/m²), Cl-pipe support coefficient, T-Time (s), Tp-pipe thickness (m), L
= distance(m).

The pressure wave generated by a flow-control operation propagated with
speed a, reaching the other end of the pipeline in a time interval equal to L/a (s
econds). The same time interval was necessary for the reflected wave to travel
back to its origin, for a total of 2 L/a (seconds). The quantity 2 L/a, termed the
characteristic time for the pipeline. It was used to classify the relative speed of a
maneuver that caused a hydraulic transient.

The effects of transients were most likely to result in pipe failures in long
pipelines with long characteristic times (large values of 2 L/a), high velocities,
and few branches. Filion and Karney [36] found that water usage and leaks in a
distribution system could result in a dramatic decay in the magnitude of transient
pressure effects [36-49].

3.2.1   PRESENT RESEARCH MODELING CAPABILITIES

The differences between computer model results and actual system measurements were caused by several factors, including the following difficulties:

- Precise determination of the pressure wave speed for the piping system was difficult, if not impossible. This was especially true for buried pipelines, whose wave speeds were influenced by bedding conditions and the compaction of the surrounding soil.
- Precise modeling of dynamic system elements (such as valves, pumps, and protection devices) was difficult because they were subject to deterioration with age and adjustments made during maintenance activities. Measurement equipment may also be inaccurate.
- Unsteady or transient friction coefficients and losses depend on fluid velocities and accelerations. These were difficult to predict and calibrate even in laboratory conditions.
- Prediction of the presence of free gases in the system liquid was sometimes impossible.

These gases can significantly affect the pressurewave speed (Fig. 1-26). In addition, the exact timing of vapor-pocket formation and column separation were difficult to simulate. Calibrating model parameters based on field data minimized the first source of error [50-59].

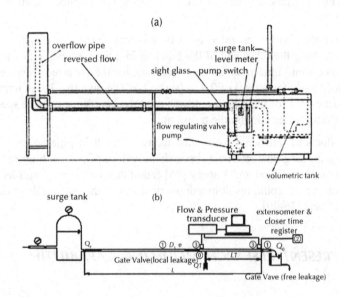

FIGURE 1 Scheme of the laboratory model: (a) present work, (b) Kodura and Weinerowska.

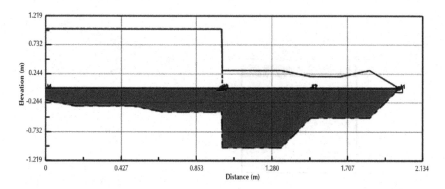

FIGURE 2   Laboratory model results (invented by: Kaveh Hariri Asli).

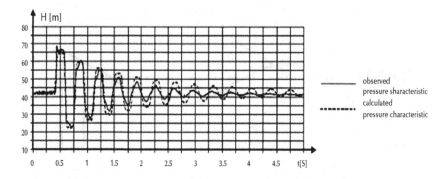

FIGURE 3   Laboratory model results of other experts' researches: Kodura and Weinerowska research.

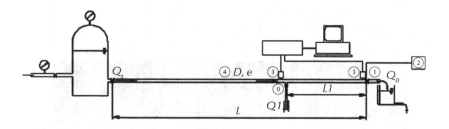

FIGURE 4   Laboratory model: Kodura and Weinerowska research.

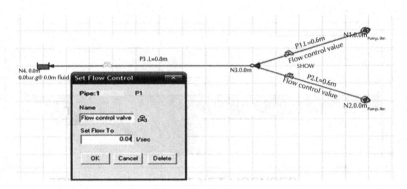

FIGURE 5   Present research laboratory model.

$$p = p_0$$

FIGURE 6   Present research laboratory model for Re. No.

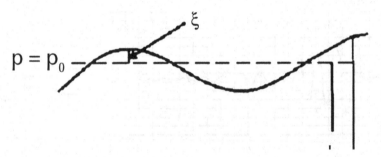

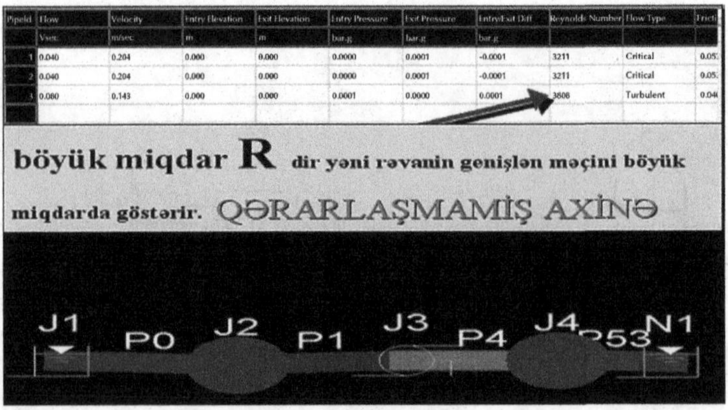

Pipeld	Flow	Velocity	Entry Elevation	Exit Elevation	Entry Pressure	Exit Pressure	Entry/Exit Diff	Reynolds Number	Flow Type	Frict
	l/sec	m/sec	m	m	bar.g	bar.g	bar.g			
1	0.040	0.204	0.000	0.000	0.0000	0.0001	-0.0001	3211	Critical	0.05
2	0.040	0.204	0.000	0.000	0.0000	0.0001	-0.0001	3211	Critical	0.05
3	0.080	0.143	0.000	0.000	0.0001	0.0000	0.0001	3600	Turbulent	0.04

böyük miqdar R dir yəni rəvanin genişlən məçini böyük miqdarda göstərir. **QƏRARLAŞMAMIŞ AXINƏ**

J1   P0   J2   P1   J3   P4   J4   253   N1

FIGURE 7   Re. No. due to fluid interpenetration.

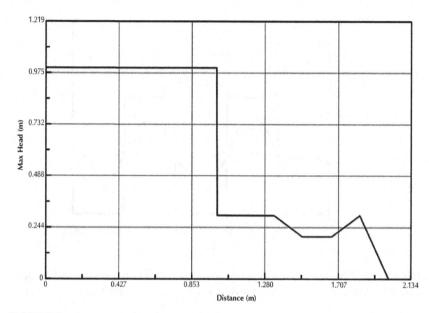

FIGURE 8   Max pressure changes due to elevation changes for water pipeline.

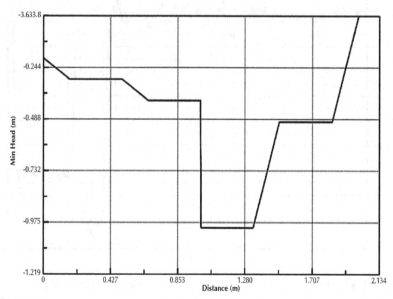

FIGURE 9   Min pressure changes due to elevation changes for water pipeline.

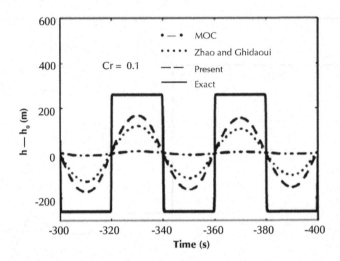

FIGURE 10   Analysis and Comparison results of calculations of other experts
Research; Pressure traces at downstream valve: Ghidaoui and Zhao (Cr = 0.1).

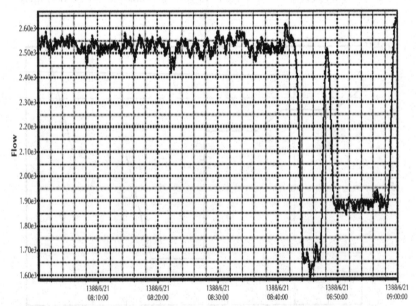

FIGURE 11   Flow changes due to time for water pipeline (first record).

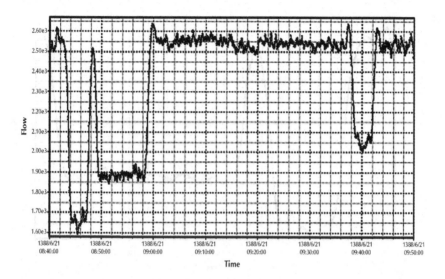

FIGURE 12   Flow changes due to time for water pipeline (second record).

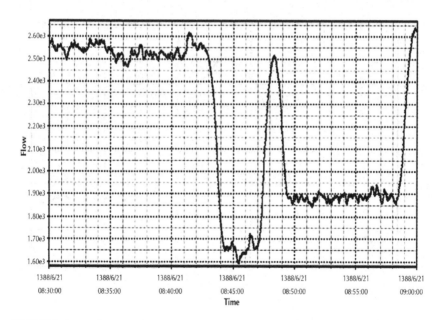

FIGURE 13   Flow changes due to time for water pipeline (third record).

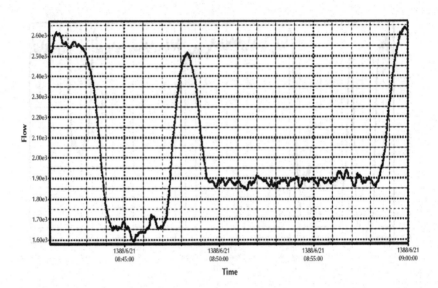

FIGURE 14   Flow changes due to time for water pipeline (forth record).

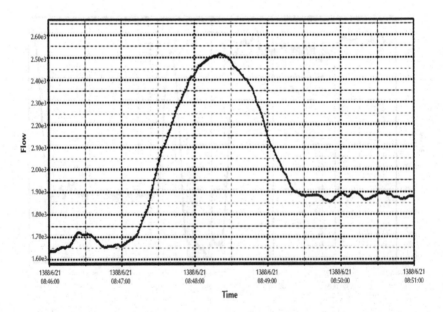

FIGURE 15   Flow changes due to time for water pipeline (fifth record).

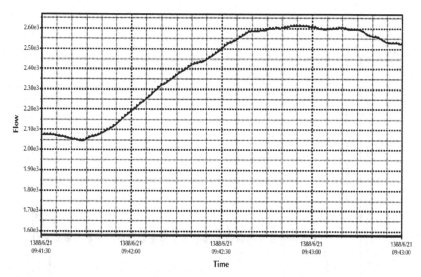

FIGURE 16   Flow changes due to time for water pipeline (sixth record).

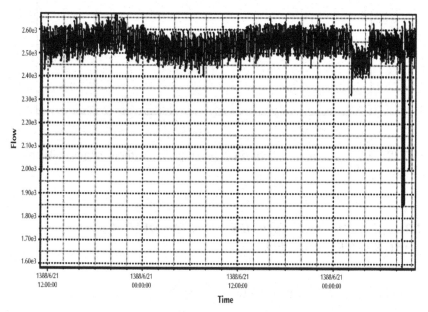

FIGURE 17   Flow changes due to time for water pipeline (seventh record).

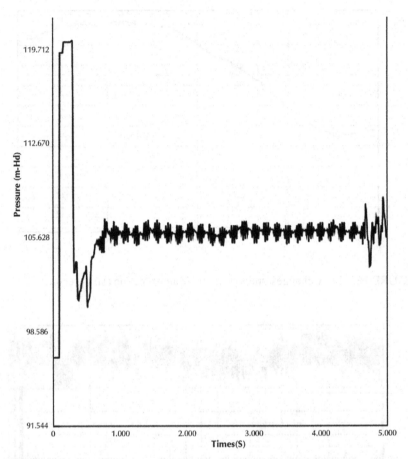

FIGURE 18 Pressure variation in water pipeline without surge tank and in leakage condition.

3.2.2 MODEL APPROACHES TO TRANSIENT FLOW (LABORATORY MODEL)

Experiments were carried out in two places:

First, by the modellocated in researcher's office laboratory, and second, model was calibratedby the model which located in hydraulic laboratory. The researcher's office laboratory modelis the First Model.

Then it was compared by laboratory model of Kodura and Weinerowska and other experts' researches. The main element of experiment instrument was the pipeline, single straight pipe of the L (length), extrinsic D (diameter) and the e (wall thickness) or the pipeline consisted of sections of varied parameters. The pipeline was equipped with the valve at the end of the main pipe, which was joined with the closure time register. The water hammer pressure characteristics were measured by extensometers, and recorded in memory of computer. The supply of the water to the system was realized with use of reservoir, which enabled inlet pressure stabilization. The experiments were carried out for cases:

- Simple positive water hammer for the straight pipeline of constant diameter; the measured characteristics were the basis for estimation the influence of the diameter change and water hammer run[60-68].

- Positive water hammer in pipeline, with the possibility of sucking in air in negative phase. This was the reason for the sucking air in negative phase for water pipeline. Consistence between observed values of maximal pressure in first.

- Significant influence of the rate of the discharge on the decreased the duration of the water hammer phenomenon. The duration time decreased with the increase of the outflow. This was the strong reason for the high deceasing in duration time for water pipeline.

3.3 RESULTS AND DISCUSSION

Present work algorithms grew and evolved to keep pace with the state of the practice in water transmission modeling. Because the mathematical solution methods are continually extended, this work deals primarily with the fundamental principles underlying these algorithms and focuses less on the details of their implementation. It also introduced the principles of hydraulic transients in piping systems, reviewed current analytical approaches and engineering practices, discussed the potential sources and impacts of water hammer, andpresented a proven approach to help for selecting and sizing surge-control equipment[69-80].

Transient simulations were integrated into the discussion to provided context. Developing a surge-control strategy-ideally, a system was designed and operated to minimize the likelihood of damaging transient events. However, in reality, transients occurred; thus, methods for controlling transients were necessary.

This work had two goals:

1. To make the hydraulic engineer aware of the system conditions that lead to the development of undesirable transients, such as pump and valve operations, and

2. To present the protection methods and devices that should be used during design and construction of particular systems and discuss their practical limitations.

There were two possible strategies for controlling transient pressures. The first was to focus on minimizing the possibility of transient conditions during project design by specifying appropriate flow-control operations and avoiding the occurrence of emergency and unusual system operations. The second was to install protection devices to control potential transients due to uncontrollable events, such as power and equipment failures. Systems protected by adequately designed surge tanks are generally not adversely affected by emergency or unusual flow-control operations, because operational failure of surge tanks is unlikely.

System was protected by gas vessels, however, an air outflow or air-compressor failure can lead to damage from transients. Consequently, potential emergency situations and failures should be evaluated and avoided to the extent possible through the use of alarms that detect device failures and control systems witch act to prevent them. With most small, well-supported water transient, sufficient safety factors were built into the system, such as adequate pipe-wall thickness and sufficient reflections (tanks and dead ends) and withdrawals (water use).

The concentrated vaporous cavity model produces satisfactory results in slow transients, but produces unstable solutions for rapid transients, such as the pump's stoppage with reflux valve closure. The discrete air release model produces satisfactory results in pump shut downcases, but is susceptible to long-term numerical damping. Typically, in the discrete air release model, the wave speed distribution along with a pipeline [81-90].

3.3.1 COMPARISON OF PRESENT WORK RESULTS WITH OTHER EXPERT'S RESEARCH

Comparison of present work results (nonlinear heterogeneous model-water hammer software modeling), with the results of other expert's works on laboratory experiments and field test results, shows similarity and advantages.

3.3.2   APOLONIUSZ KODURA AND KATARZYNA WEINEROWSKA, (2005)

In present work water hammer has been run in pressurized pipeline. The observations, experiments and numerical analysis shows air existence in the pipeline have generated the complex condition for water hammer phenomenon. Therefore, it has been influenced by some additional factors. The effect of total discharge of the pipeline on the periods of wave oscillations has been investigated. The pipeline was equipped with the valve at the end of the main pipe, which was joined with the closure time register. The water hammer pressure characteristics were measured by extensometers, and recorded in computer memory. The supply of the water to the system was realized with use of reservoir, which enabled inlet pressure stabilization.

The pressure wave speed was a fundamental parameter for hydraulic transient modeling at present research, since it determined how quickly disturbances propagate throughout the system. This affected whether or not different pulses may superpose or cancel each other as they meet at different times and locations. Wave speed was affected by pipe material and bedding, as well as by the presence of fine air bubbles in the fluid.

The pressure wave generated by a flow-control operation propagated with speed a, reaching the other end of the pipeline in a time interval equal to L/a (seconds). The same time interval was necessary for the reflected wave to travel back to its origin, for a total of

$2\ L/a$ (seconds). The quantity $2\ L/a$, termed the characteristic time for the pipeline. It was used to classify the relative speed of a maneuver that caused a hydraulic transient.

Unsteady or transient friction coefficients and the effects of free gases are more challenging to account for. Fortunately, friction effects are usually minor in most water systems and vaporization can be avoided by specifying protection devices and/or stronger pipes and fittings able to withstand sub atmospheric or vacuum conditions, which are usually short-lived.

For present research with free gas systems and the potential for water-column separation, the numerical simulation of hydraulic transients was more complex. Small pressure spikes caused by the type of tiny vapor pockets that was difficult to simulate accurately seldom result in a significant change to the transient envelops. Larger vapor-pocket collapse events resulting in significant upsurge pressures were simulated with enough accuracy to support definitive conclusions.

In the current research water hammer modeling has been analyzed in two manners:

1. No leakage assumption for transmission line.
2. No leakage assumption for transmission line witch equipped by pressure vessel or two way surge tank (initial condition).

3.3.3 ADVANTAGE OF PRESENT RESEARCH, COMPARE WITH OTHER EXPERT'S RESEARCH

1. Data gathering prepared a modeling regression line with acceptable low coefficient of correlation (r). Soresearch-modeling functionhad anacceptable correlation around the scatter diagram.
2. Observation and record of gages and "PLC" or "SCADA" systems, high-speed sensors and data logger equipment's output values that installed to accurately track transient eventson water Transmission Line.
3. Similar to other experts, present researchfor water transmission line simulationhas been used water hammer software (MOC).
4. The specific and main advantage of present research in comparison with all of other expert's research in the world. Therefore it proved advantage of present research method compared with other expert's research results. Hence it was compared with other expert's research results, such as:

3.3.4 COMPARISON OF PRESENT RESEARCH WITH OTHER EXPERT'S RESEARCH

Comparison of present research results(water hammer modeling by "MOC" and SPSS modeling software), with other expert's research results, showed too similarity and some advantages:

ANTON BERGANT [22]

Time adjustments for cavity opening and collapse were implemented into the classical discrete vapor cavity model (DVCM) with steady pipe flow friction and their influence on pressure spikes and the time of cavity existence was investigated. In the current research sub-atmospheric or even full-vacuum pressures. The vapor pocket caused a dramatic high-pressure transient when the water column rejoins very rapidly, Analysis showed sub atmospheric or even full-vacuum pressures was generated in the near of reservoir. Thus a little air interred into pipeline, which it must be removed from the system.

Results without time adjustments of cavity opening and collapse were presented as a baseline solution and compared with measured pressure histories. The simulation results show that adjustment of the timing of the cavity collapse has a greater influence on pressure pulses than the time adjustment for cavity opening. Although the time adjustment of cavity collapse produces some larger pressure spikes, a much better timing of the transient event is obtained compared to results of classical DVCM and of DVCM with adjusted cavity opening. The discrepancies between the different computed results found by time-history comparisons may be attributed to the intensity of cavitations along the pipeline (distributed vaporous cavitations regions, actual number and position of intermediate cavities) resulting in a slightly different timing of cavity collapse and a different superposition of waves. The obtained improvements to DVCM are small but important, because the method is so much used in practice.

BRUNONE ET AL. [24]

In spite of a big progress in mathematical modeling and measurements, water hammer is still one of the most interesting problems of pipeline hydraulics, and the subjects of numerous publications, as one of those problems, which still are not recognized in sufficient way. Present research aimed at the definition of practical manifest for dealing with fluid-structure interaction "FSI" in water pipeline systems. For unsteady flow in pipes, this leads to a natural set of regimes. One of these regimes was water hammer with FSI; this was the regime where pipe motion starts to influence the unsteady flow (earthquake condition). The main objective of this research was to identify this particular condition with flows in pipes that were not fully restrained (i.e., were able to move in response to internal fluid forces).

WYLIE, E.B. AND STREETER, V.L. [148]

Classical water hammer theory work of Wylie, E. B., and Streeter, V. L. [148] neglects convective terms, and assumes fluid wave speed, c(m/s) is depend on the support conditions of the pipes.

At water hammer and 1D – FSI Fluid when sufficiently rapid accelerations occur, fluid compressibility effects come into play and it ceases to be acceptable to neglect nonuniformities in velocity along any particular pipe. It becomes necessary to take explicit account of pressure waves that propagate at the speed of sound c in the liquid.

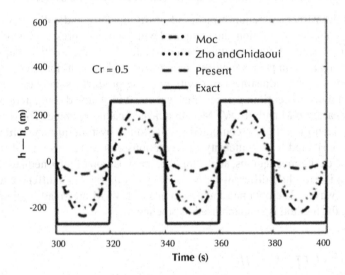

FIGURE 19 Analysis and Comparison results of calculations of other experts Research; Pressure traces at downstream valve: Ghidaoui and Zhao (Cr = 0.5).

ARRIS S. TIJSSELING AND ALAN E. VARDY[150]

Present research assumed four cases in field tests; transmission line with and without surge tank. So present research results:have been compared with workof Arris S. Tijsseling, and Alan E. VardyResearch results. Comparison showed similarity in results. Their paper was the first attempt to define time scales for problems possibly involving fluid-structure interaction"FSI."

Their equations were a set of hyperbolic partial differential equations. By using the Reynolds transport theorem and assuming one-dimensional flow and elastic conduit with slightly compressible fluid, which stretch or contracts with respect to time.

CHAUDHRY[29] AND JAIME SUAREZ ACUNA [106]

Obtained pressure heads by the steady and unsteady friction model. Comparison showed similarity In the current research and workof Chaudhry, [29] and Jaime Suárez Acuña [106] research results.

GHIDAOUI ET AL., 2005

In fact, in a review of commercially available water hammer software packages, it is found that 11 out of 14 software packages examined use MOC schemes.

CHAUDHRY AND HUSSAINI 1985, AND WOOD ET AL. [161]

Among the approaches proposed to solve the one-phase (pure liquid) water hammer equations are the Method of Characteristics (MOC), Finite Differences (FD), Wave Characteristic Method (WCM), Finite Elements (FE), and Finite Volume (FV). In-depth discussions of these methods can be found in work of Chaudhry and Hussaini 1985, and Wood et al. [161].

PARMAKIAN, 1955, AND WYLIE AND STREETER [148]

Among these methods, MOC-based schemes were most popular because these schemes provided the desirable attributes of accuracy, numerical efficiency and programming simplicity in the workof Parmakian, 1955, and Wylie and Streeter [148]. Thewater hammer software packages, which were used in the current research, applied Numerical solutions of the nonlinear Navier-Stokes equations by Method of Characteristics (MOC).

WYLIE AND STREETER [148] AND CHAUDHRY [29]

The partial differential equations that describe two-phase flows in closed conduits can be simplified to a great extent when the amount of gas in the conduit is small. In this case, the gas-liquid mixture can be treated as a single-equivalent fluid in work of Wylie and Streeter [148], and Chaudhry [29]. For the present research with free gas systems and the potential for water-column separation, the numerical simulation of hydraulic transients was used.

MARTIN 1993, AND WYLIE AND STREETER [148]

Research focused on the formulation and numerical efficiency assessment of a second-order accurate FV shock-capturing scheme for simulating one and two-phase water hammer flows. In this case, it was assumed that there was no relative motion or slip between the gas and the liquid and both phases. It treated as a single-equivalent fluid with average properties work of Martin 1993, and Wylie and Streeter [148]. For Sudden changes, estimating the exact timing of vapor-pocket formation and column separation were difficult to simulate. Calibrating model parameters based on field data minimized the source of error.

ARTURO S. LEON[116]

Comparison showed similarity between present research results against workof Arturo S. Leon, 2007, research results were as follows:
- Numerical tests showed that the proposed second-order formulation at boundary conditions (achieved by using virtual cells) is second-order. In addition, the proposed formulation maintains the conservation property of FV schemes and introduced no unphysical perturbations into the computational domain.
- Numerical tests were performed for smooth (i.e., flows that do not presentdiscontinuities) and sharp transients. The results show that the efficiency of the proposed scheme is superior to both the MOC scheme and the second-order FV scheme of Zhao and Ghidaoui.
- The high efficiency of the proposed scheme was important for Real-Time Control RTC of water hammer flows in large networks.

In the current research, "Interpenetration of two fluids at parallel between plates and turbulent moving in pipe" Changes at system boundaries (Sudden changes) created a transient pressure pulse. In this regard, model design needed to find the relation between many variables accordance to fluid transient.Therefore, a computational technique has been presented and the results have been compared by field tests.

APOLONIUSZ KODURA [19] AND KATARZYNA WEINEROWSKA [109]

Research presented water hammer run in pressure pipeline. The results of experiments and numerical analysis of the phenomenon were presented. The phenomenon run was influenced by some additional factors. Detailed conclusions drawn on the basis of experiments and calculations for the pipeline presented in the paper of Kodura and Weinerowska [109].

The most important effects observed were the Experiments were carried out in the laboratory of Warsa, University of Technology, Environmental Engineering Faculty, Institute of Water Supply and Water Engineering. In the current research, Experiments were carried out by the model located in researcher's office laboratory [91-149].

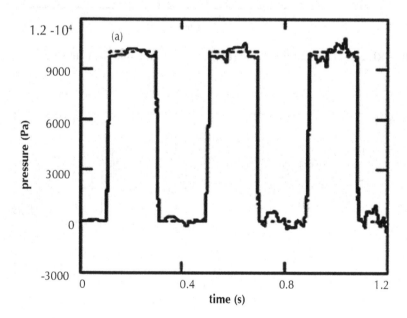

FIGURE 20   *(Contitued)*

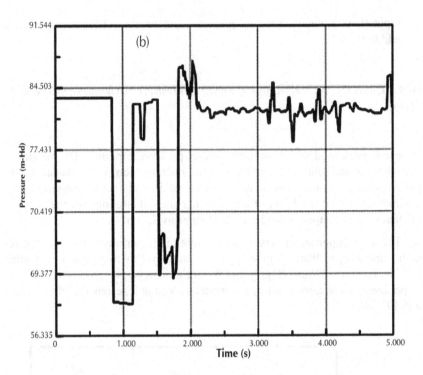

FIGURE 20 Transmission Line: (a) pressure at midpoint (b) with surge tank.

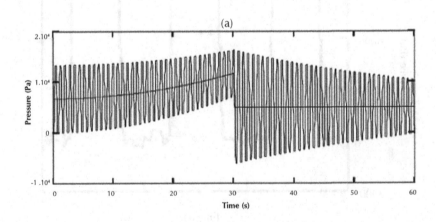

FIGURE 21 *(Continued)*

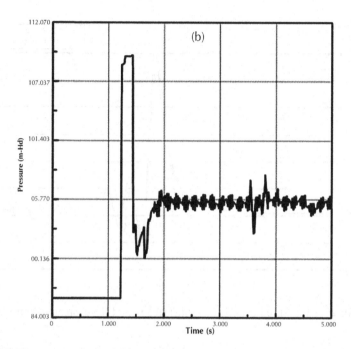

FIGURE 21 Transmission Line: (a) Starting flow driven by 2 (m/s) velocity rise in 30 seconds time, (b) without surge tank.

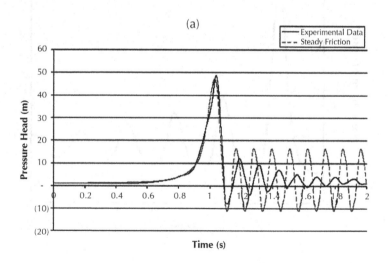

FIGURE 22 *(Continued)*

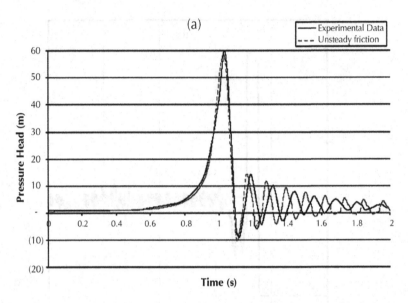

FIGURE 22   Pressure head histories: (a) steady and (b) unsteady friction.

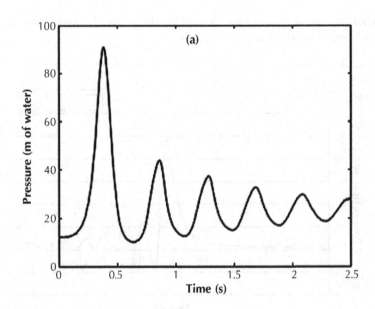

FIGURE 23   *(Contitued)*

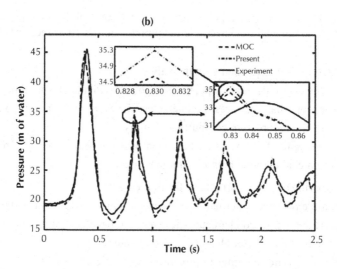

FIGURE 23 Absolute pressure: (a) Experimental and (b) MOC method.

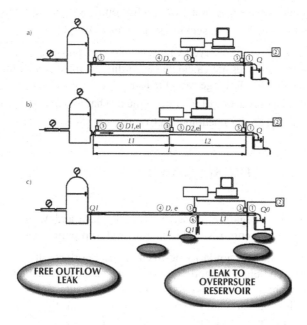

FIGURE 24 Experimental equipment (a) simple pipeline, (b) pipeline with diameter change, (c) pipeline.

The main element is the pipeline (single straight pipe of the length L, extrinsic diameter D and the wall thickness e or the pipeline consisted of sections of varied parameters).

The pipeline was equipped with the valve:

- At the end of the main pipe, this was joined with the closure time register.
- The water hammer pressure characteristics were measured by extensometers and recorded in computer's memory.
- The supply of the water to the system was realized with use of reservoir, which enabled inlet pressure stabilization.

The experiments were carried out for four cases:

- Simple positive water hammer for the straight pipeline of constant diameter; the measured characteristics were the basis for estimation the influence of the diameter change and water hammer run:
- Positive water hammer in pipeline with single change of diameter: contraction and extension.
- Positive water hammer in pipeline in two scenarios: with the outflow to the overpressure reservoir and with free outflow (to atmospheric pressure, with the possibility of sucking in air in negative phase).

Consistence between observed values of maximal pressure in first amplitude and corresponding values are calculated according to Joukowski's formula, irrespective of the rate of discharge from the leak;

- Significant influence of the rate of the discharge decreased of duration of the water hammer phenomenon Comparison shows similarity in results.

KODURA AND WEINEROWSKA [19]

For sucking air in negative phase, work of Kodura and Weinerowska [19], experimental equipment: a) simple pipeline and b) pipeline with flow change.

Comparison showed similarity between present work results water transmission line (research field tests model: Elevation-Distance transient curve for transmission line without surge tank– Max. transient pressure line – Min. pressure line – steady flow pressure line) with work of Kodura and Weinerowska. There is no minus pressure and air volume in a zone of transmission line. But in no leakage condition and sucking in air in negative phase, (research field tests model: Elevation-Distance transient curve for transmission line with surge tank and no leakage condition). There is minus pressure in a zone of transmission line, in the near of water reservoir, so was generated a little air volume so it must be removed from the system. Experimental equipment in the work of Kodura and Weinerowska [19], has not been equipped with advanced flow and pressure sensors coupled with high-speed data loggers which make it possible to capture fast transients, down to 5 milliseconds for interpenetration between water flows. On the other hand, it was a scale model, which could be built to reproduce transients observed in a prototype (real) system, for comparing by field tests [42-61].

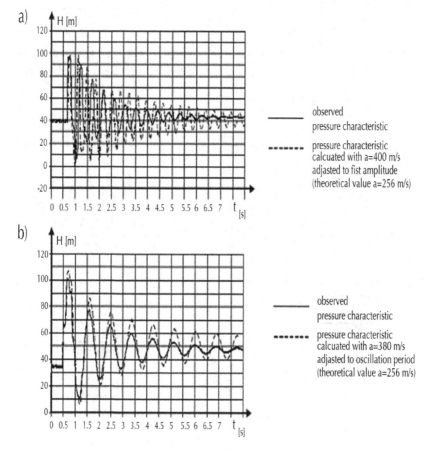

FIGURE 25 Pipeline with change of diameter; work of Kodura and Weinerowska: (a) contraction, (b) extension.

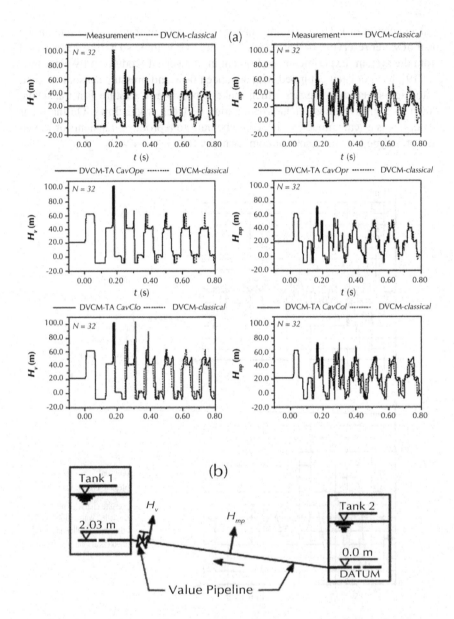

FIGURE 26 (a) Comparison of heads at the valve and at the midpoint, (b) Experimental apparatus layout.

ARTURO S. LEON [116]

Work of Arturo S. Leon, comparison requires measuring the CPU time needed by each of the schemes to achieve the same level of accuracy work of (i.e., Zhao).

GHIDAOUI, 2004 AND LEYN ET AL. [116]

The efficiency of a model is a critical factor for Real-Time Control (RTC), since several simulations are required within a control loop in order to optimize the control strategy, and small simulation time steps are needed to reproduce the rapidly varying hydraulics. Comparison shows similarity in work of León et al. [116], andpresent work results.

3.3.5   THE MAXIMUM VALUE OF PRESSURE

This work led to improved standards for precession designs and installation techniques in the field of subatmospheric transient pressures.

A treated or modeled air entrainment problem in real prototype systems and results was showed. Consistency between the observed values of maximal pressure in the first amplitude and corresponding values were calculated according to Joukowski's formula, irrespective of the rate of flow. Significant influence of the discharge rate into the pipeline decreases of duration of the water hammer phenomenon. The duration time decreased with the increase of the air penetration. This was the strong reason for the high deceasing in duration time for water pipeline. In this work the column separations due to the turned off pump for the water pipeline and the air penetration were carried out. Pressure at the beginning of pipeline remains constant. After shock wave generation, flow moves with the same velocity at the reverse direction of the shock wave.

This leads to the generation of high-pressure drop due to the wave. At the same time fluid moves in the reverse direction of the pipeline. As long as the shock wave reaches to the pressure reducing valve, liquid pressure reduces to vapor pressure. By the way, again and again, the wave of pressure drop moves conversely in the direction of start point on the water pipeline. As long as damping of shock wave, these cycles of increase and decrease of pressure will be continued. It is iterated at time intervals equal to time for dual-path of the shock wave along with the length of the pipeline (from the pressure reducing valve prior to the start point of pipeline). The hydraulic impact of the liquid in the pipeline will perform oscillatory motion. The hydraulic resistance and viscosity cause the oscillatory motion. It absorbs the initial energy of the liquid as long as overcoming the friction and therefore it will be damped. Water hammer is manifested in hydro-machines various purposes. In most cases this is undesirable, leading to the destruction of pipelines [150-170].

Maximum amount of air infiltration, which was calculated based on the simulation results of nonlinear heterogeneous model were released by the air relief valve on the system.

CONCLUSIONS

The water hammer pressure characteristics were measured by extensometers, and recorded in computer's memory. The supply of the water to the system was realized with use of reservoir, which enabled inlet pressure stabilization.

KEYWORDS

- **Computational method**
- **Hydraulic transient modeling**
- **Hydraulic transition**
- **Surge tank**
- **Water hammer**

REFERENCES

1. Hariri Asli, K.; GIS Water hammer disaster at earthquake in Rasht water pipeline, 3rd International Conference on Integrated Natural Disaster Management, INDM **2008,** http://www.civilica.com/Paper-INDM03-INDM03_001.html.
2. Hariri Asli, K.; Nagiyev, F. B.; Beglou, M. J.; Haghi, A. K.; Kinetic analysis of convective drying, International Journal of the Balkan Tribological Association, ISSN: 1310–4772, Sofia, Bulgaria, **2009,** *15(4),* 546–556, jbalkta@gmail.com

3. Hariri Asli, K.; Nagiyev, F. B.; Bubbles characteristics and convective effects in the binary mixtures. Transactions issue mathematics and mechanics series of physical-technical and mathematics science, ISSN 0002–3108, Azerbaijan, Baku, 215–220, **2008,** www.imm.science.az/journals/AMEA_xeberleri/.../215–220.pdf

4. Hariri Asli, K.; Nagiyev, F. B.; Haghi, A. K.; Three-dimensional Conjugate Heat Transfer in Porous Media, International Journal of the Balkan Tribological Association, ISSN: 1310–4772, Sofia, Bulgaria, **2009,** *15(3),* 336–346, jbalkta@gmail.com

5. Hariri Asli, K.; Nagiyev, F. B.; Haghi, A. K.; Water hammer and fluid condition; a computational approach, Computational Methods in Applied Science and Engineering, USA, Chapter *5,* Nova Science Publications, ISBN: 978-1-60876-052-7, USA, **2010,** 73–94, https://www.novapublishers.com/catalog/

6. Hariri Asli, K.; Nagiyev, F. B.; Haghi, A. K.; Interpenetration of two fluids at parallel between plates and turbulent moving in pipe; a case study, Computational Methods in Applied Science and Engineering, USA, Chapter 7, Nova Science Publications, ISBN: 978-1-60876-052-7, USA, **2010,** 107–133, https://www.novapublishers.com/catalog/

7. Hariri Asli, K.; Nagiyev, F. B.; Haghi, A. K.; Modeling for water hammer due to valves; from theory to practice, Computational Methods in Applied Science and Engineering, USA, Chapter 11, Nova Science Publications ISBN: 978-1-60876-052-7, USA, **2010,** 229–236, https://www.novapublishers.com/catalog/

8. Hariri Asli, K.; Nagiyev, F. B.; Haghi, A. K.; A computational method to Study transient flow in binary mixtures, Computational Methods in Applied Science and Engineering, USA, Chapter 13, Nova Science Publications ISBN: 978-1-60876-052-7, USA, **2010,** 229–236, https://www.novapublishers.com/catalog/

9. Hariri Asli, K.; Nagiyev, F. B.; Haghi, A. K.; Water hammer analysis; some computational aspects and practical hints, Computational Methods in Applied Science and Engineering, USA, Chapter 16, Nova Science Publications ISBN: 978-1-60876-052-7, USA, **2010,** 263–281, https://www.novapublishers.com/catalog/

10. Hariri Asli, K.; Nagiyev, F. B.; Haghi, A. K.; Water hammer and hydrodynamics instabilities modeling, Computational Methods in Applied Science and Engineering, USA, Chapter 17, From Theory to Practice, Nova Science Publications ISBN: 978-1-60876-052-7, USA, **2010,** 283–301, https://www.novapublishers.com/catalog/

11. Hariri Asli, K.; Nagiyev, F. B.; Haghi, A. K.; A computational approach to study water hammer and pump pulsation phenomena, Computational Methods in Applied Science and Engineering, USA, Chapter 22, Nova Science Publications, ISBN: 978-1-60876-052-7, USA, **2010,** 349–363, https://www.novapublishers.com/catalog/

12. Hariri Asli, K.; Nagiyev, F. B.; Haghi, A. K.; Some aspects of physical and numerical modeling of water hammer in pipelines. Computational Methods in Applied Science and Engineering, USA, Chapter 23, Nova Science Publications, ISBN: 978-1-60876-052-7, USA, **2010,** 365–387, https://www.novapublishers.com/catalog/

13. Hariri Asli, K.; Nagiyev, F. B.; Haghi, A. K.; A computational approach to study fluid movement, Nanomaterials Yearbook – **2009,** From Nanostructures, Nanomaterials and Nanotechnologies to Nanoindustry, Chapter 16, Nova Science Publications, USA, ISBN: 978-1-60876-451-8, USA, **2010,** 181–196. https://www.novapublishers.com/catalog/product_info.php?products_id=11587

14. Hariri Asli, K.; Nagiyev, F. B.; Haghi, A. K.; Physical modeling of fluid movement in pipelines, Nanomaterials Yearbook – 2009, From Nanostructures, Nanomaterials and Nanotechnologies to Nanoindustry, Chapter 17, Nova Science Publications, USA, ISBN: 978-1-60876-451-8, USA, 2010, 197–214. https://www.novapublishers.com/catalog/product_info.php?products_id=11587

15. Hariri Asli, K.; Nagiyev, F. B.; Haghi, A. K.; Aliyev, S. A.; Improved modeling for prediction of water transmission failure, Recent Progress in Research in Chemistry and Chemical Engineering, Chapter 2, Nova Science Publications, ISBN: 978-1-61668-501-0, Nova Science Publications, USA, 28–36, 2010, https://www.novapublishers.com/catalog/product_info.php?products_id=13174

16. Hariri Asli, K.; Nagiyev, F. B.; Haghi, A. K.; Aliyev S. A.; Pure Oxygen penetration in wastewater flow, Recent Progress in Research in Chemistry and Chemical Engineering, Chapter 3, Nova Science Publications, ISBN: 978-1-61668-501-0, Nova Science Publications, USA, 2010, 17–27, https://www.novapublishers.com/catalog/product_info.php? products_id=13174

17. Adams, T. M.; Abdel-Khalik, S. I.; Jeter, S. M.; Qureshi, Z. H. An Experimental investigation of single-phase Forced Convection in Microchannels, *Int.J. Heat Mass Transfer,* 1998, *41,* 851–857.

18. Allievi L. "General Theory of Pressure Variation in Pipes,"*Ann. D.; Ing.* 1982, 166–171.

19. Apoloniusz Kodura, Katarzyna Weinerowska," Some Aspects of Physical and Numerical Modeling of Water Hammer in Pipelines," 2005, 126–132.

20. Anuchina, N. N.; Volkov V. I.; Gordeychuk V. A.; Es'kov, N. S.; Ilyutina, O. S.; Kozyrev O. M. "Numerical simulations of Rayleigh-Taylor and Richtmyer-Meshkov instability using mah-3 code,"*J. Comput. Appl. Math.*2004, *168,* 11.

21. Bergeron L. "Water hammer in Hydraulics and Wave Surge in Electricity," John Wiley and Sons, Inc.; N.Y. 1961, 102–109.

22. Bergant Anton Discrete Vapour Cavity Model with Improved Timing of Opening and Collapse of Cavities, 1980, 1–11.

23. Bracco A.; McWilliams, J. C.; Murante G.; Provenzale A.; Weiss, J. B.; "Revisiting freely decaying two-dimensional turbulence at millennial resolution," Phys. Fluids, Issue2000, *11, 12,* 2931–2941.

24. Brunone B.; Karney, B. W.; Mecarelli M.; Ferrante M. "Velocity Profiles and Unsteady Pipe Friction in Transient Flow" Journal of Water Resources Planning and Management, ASCE, Jul. 2000, *126(4),* 236–244.

25. Cabot, W. H.; Cook, A. W.; Miller, P. L.; Laney, D. E.; Miller, M. C.; Childs, H. R.; "Large eddy simulation of Rayleigh-Taylor instability," Phys. Fluids, September, 2005, *17,* 91–106.

26. Cabot W.; University of California, Lawrence Livermore National laboratory, Livermore, CA, *Phys. Fluids,* 2006, 94–550.

27. Chaudhry M. H. "Applied Hydraulic Transients," Van Nostrand Reinhold Co. N.Y. 1979, 1322–1324.

28. Chaudhry, M. H.; Yevjevich, V. "Closed Conduit Flow," Water Resources Publication, USA, 1981, 255–278.

29. Chaudhry M. H.; Applied Hydraulic Transients, Van Nostrand Reinhold, New York, USA, 1987, 165–167.

30. Clark, T. T.; "A numerical study of the statistics of a two-dimensional Rayleigh-Taylor mixing layer,"*Phys. Fluids* **2003**, *15*, 2413p.
31. Cook, A. W.; Cabot, W.; Miller, P. L.; "The mixing transition in Rayleigh-Taylor instability,"*J. Fluid Mech.***2004**, *511*, 333 p.
32. Choi, S. B.; Barren, R. R.; Warrington, R. O.; Fluid Flow and Heat Transfer in Microtubes, *ASME DSC* **1991**, *40*, 89–93.
33. Dimonte, G.; Youngs, D.; Dimits, A.; Weber, S.; Marinak, M. "A comparative study of the Rayleigh-Taylor instability using high-resolution three-dimensional numerical simulations: the Alpha group collaboration,"*Phys. Fluids* **2004**, *16*, **1668**,
34. Dimotakis, P. E.; "The mixing transition in turbulence,"*J. Fluid Mech.***2000**, *409*, 69 p.
35. Elansari, A. S.; Silva , W.; Chaudhry M. H. "Numerical and Experimental Investigation of Transient Pipe Flow," Journal of Hydraulic Research, **1994**, *32*, 689 p.
36. Filion, Y.; Karney, B. W.; "A Numerical Exploration of Transient Decay Mechanisms in Water Distribution Systems," Proceedings of the ASCE Environmental Water Resources Institute Conference, American Society of Civil Engineers, Roanoke, Virginia, **2002**, 30 p.
37. Fok, A.; "Design Charts for Air Chamber on Pump Pipelines,"*J. Hyd. Div. ASCE*, Sept.; **1978**, 15–74.
38. Fok, A.; Ashamalla A.; Aldworth G.; "Considerations in Optimizing Air Chamber for Pumping Plants," Symposium on Fluid Transients and Acoustics in the Power Industry, San Francisco, USA, Dec, **1978**, 112–114.
39. Fok, A.; "Design Charts for Surge Tanks on Pump Discharge Lines," BHRA 3rd Int. Conference on Pressure Surges, Bedford, England, Mar.; **1980**, 23–34.
40. Fok, A.; "Water hammer and Its Protection in Pumping Systems," Hydro technical Conference, CSCE, Edmonton, May, **1982**, 45–55.
41. Fok, A.; "A contribution to the Analysis of Energy Losses in Transient Pipe Flow," PhD.; Thesis, University of Ottawa, **1987**, 176–182.
42. Fox, J. A.; "Hydraulic Analysis of Unsteady Flow in Pipe Network," Wiley: N.Y. **1977**, 78–89.
43. Fedorov, A. G.; Viskanta, R.; Three-dimensional Conjugate Heat Transfer into Microchannel Heat Sink for Electronic Packaging, *Int.J. Heat Mass Transfer* **2000**, *43*, 399–415.
44. Hamam, M. A.; Mc Corquodale, J. A.; "Transient Conditions in the Transition from Gravity to Surcharged Sewer Flow,"*Canadian J. Civil Eng.;* Canada, Sep.; **1982**, 65–98.
45. Hariri Asli, K.; Nagiyev, F. B.; Water Hammer and fluid condition, Ministry of Energy, Gilan Water and Wastewater Co.; Research Week Exhibition, Tehran, Iran, December, **2007**, 132–148, http://isrc.nww.co.ir.
46. Hariri Asli, K.; Nagiyev, F. B.; Water Hammer analysis and formulation, Ministry of Energy, Gilan Water and Wastewater Co.; Research Week Exhibition, Tehran, Iran, December, **2007**, 111–131, http://isrc.nww.co.ir.
47. Hariri Asli, K.; Nagiyev, F. B.; Water Hammer and hydrodynamics instabilities, Interpenetration of two fluids at parallel between plates and turbulent moving in pipe, Ministry of Energy, Guilan Water and Wastewater Co.; Research Week Exhibition, Tehran, Iran, December, **2007**, 90–110, http://isrc.nww.co.ir.

48. Hariri Asli, K.; Nagiyev, F. B.; Water Hammer and pump pulsation, Ministry of Energy, Guilan Water and Wastewater Co.; Research Week Exhibition, Tehran, Iran, December, **2007**, 51–72, http://isrc.nww.co.ir.

49. Hariri Asli, K.; Nagiyev, F. B.; Reynolds number and hydrodynamics' instability," Ministry of Energy, Guilan Water and Wastewater Co.; Research Week Exhibition, Tehran, Iran, December, **2007**, 31–50, http://isrc.nww.co.ir.

50. Hariri Asli, K.; Nagiyev, F. B.; Water Hammer and valves, Ministry of Energy, Guilan Water and Wastewater Co.; Research Week Exhibition, Tehran, Iran, December, **2007**, 20–30, http://isrc.nww.co.ir.

51. Hariri Asli, K.; Nagiyev, F. B.; "Interpenetration of two fluids at parallel between plates and turbulent moving in pipe," Ministry of Energy, Guilan Water and Wastewater Co.; Research Week Exhibition, Tehran, Iran, December, **2007**, 73–89, http://isrc.nww.co.ir.

52. Hariri Asli, K.; Nagiyev, F. B.; Decreasing of Unaccounted For Water "UFW" by Geographic Information System "GIS" in Rasht urban water system, civil engineering organization of Guilan, Technical and Art Journal, **2007**, 3–7, http://www.art-of-music.net/.

53. Hariri Asli, K.; Portable Flow meter Tester Machine Apparatus, Certificate on registration of invention, Tehran, Iran, #010757, Series a/82, 24/11/2007, 1–3.

54. Hariri Asli, K.; Nagiyev, F. B.; Haghi, A. K.; "Interpenetration of two fluids at parallel between plates and turbulent moving in pipe," 9th Conference on Ministry of Energetic works at research week, Tehran, Iran, **2008**, 73–89, http://isrc.nww.co.ir.

55. Hariri Asli, K.; Nagiyev, F. B.; Haghi, A. K.; "Water hammer and valves," 9th Conference on Ministry of Energetic works at research week, Tehran, Iran, **2008**, 20–30, http://isrc.nww.co.ir.

56. Hariri Asli, K.; Nagiyev, F. B.; Haghi, A. K.; "Water hammer and hydrodynamics instability," 9th Conference on Ministry of Energetic works at research week, Tehran, Iran, **2008**, 90–110, http://isrc.nww.co.ir.

57. Hariri Asli, K.; Nagiyev, F. B.; Haghi, A. K.; "Water hammer analysis and formulation," 9th Conference on Ministry of Energetic works at research week, Tehran, Iran, **2008**, 27–42, http://isrc.nww.co.ir.

58. Hariri Asli, K.; Nagiyev, F. B.; Haghi, A. K.; "Water hammer & fluid condition," 9th Conference on Ministry of Energetic works at research week, Tehran, Iran, **2008**, 27–43, http://isrc.nww.co.ir.

59. Hariri Asli, K.; Nagiyev, F. B.; Haghi, A. K.; "Water hammer and pump pulsation," 9th Conference on Ministry of Energetic works at research week, Tehran, Iran, **2008**, 27–44, http://isrc.nww.co.ir.

60. Hariri Asli, K.; Nagiyev, F. B.; Haghi, A. K.; "Reynolds number and hydrodynamics instability," 9th Conference on Ministry of Energetic works at research week, Tehran, Iran, **2008**, 27–45, http://isrc.nww.co.ir.

61. Hariri Asli, K.; Nagiyev, F. B.; Haghi, A. K.; "Water hammer and fluid Interpenetration," 9th Conference on Ministry of Energetic works at research week, Tehran, Iran, **2008**, 27–47, http://isrc.nww.co.ir.

62. Hariri Asli, K.; GIS and water hammer disaster at earthquake in Rasht water pipeline, civil engineering organization of Guilan, Technical and Art Journal, **2008**, 14–17, http://www.art-of-music.net/.

63. Hariri Asli, K.; GIS and water hammer disaster at earthquake in Rasht water pipeline, 3rd International Conference on Integrated Natural Disaster Management, Tehran university, ISSN: 1735–5540, 18–19 Feb.; INDM, Tehran, Iran, **2008**, *13*, 53/1–12, http://www.civilica.com/Paper-INDM03-INDM03_001.html

64. Hariri Asli, K.; Nagiyev, F. B.; Bubbles characteristics and convective effects in the binary mixtures. Transactions issue mathematics and mechanics series of physical-technical and mathematics science, ISSN: 0002–3108, Azerbaijan, Baku, **2009**, 68–74, http://www.imm.science.az/journals.html.

65. Hariri Asli, K.; Nagiyev, F. B.; Haghi, A. K.; Aliyev, S. A.; Three-Dimensional conjugate heat transfer in porous media, 1st Festival on Water and Wastewater Research and Technology, Tehran, Iran, 12–17 Dec. **2009**, 26, 28, http://isrc.nww.co.ir.

66. Hariri Asli, K.; Nagiyev, F. B.; Haghi, A. K.; Aliyev, S. A.; Some Aspects of Physical and Numerical Modeling of water hammer in pipelines, 1st Festival on Water and Wastewater Research and Technology, Tehran, Iran, 12–17 Dec. **2009**, 26, 29, http://isrc.nww.co.ir

67. Hariri Asli, K.; Nagiyev, F. B.; Haghi, A. K.; Aliyev, S. A.; Modeling for Water Hammer due to valves: From theory to practice, 1st Festival on Water and Wastewater Research and Technology, Tehran, Iran, 12–17 Dec. **2009,** 26, 30, http://isrc.nww.co.ir.

68. Hariri Asli, K.; Nagiyev, F. B.; Haghi, A. K.; Aliyev, S. A.; Water hammer and hydro-dynamics instabilities modeling: From Theory to Practice, 1st Festival on Water and Wastewater Research and Technology, Tehran, Iran, 12–17 Dec. **2009,** 26, 31, http://isrc.nww.co.ir

69. Hariri Asli, K.; Nagiyev, F. B.; Haghi, A. K.; Aliyev, S. A.; A computational approach to study fluid movement, 1st Festival on Water and Wastewater Research and Technology, Tehran, Iran, 12–17 Dec. **2009**, 27–32, http://isrc.nww.co.ir.

70. Hariri Asli, K.; Nagiyev, F. B.; Haghi, A. K.; Aliyev, S. A.; Water Hammer Analysis: Some Computational Aspects and practical hints, 1st Festival on Water and Wastewater Research and Technology, Tehran, Iran, 12–17 Dec. **2009**, 27–33, http://isrc.nww.co.ir

71. Hariri Asli, K.; Nagiyev, F. B.; Haghi, A. K.; Aliyev, S. A.; Water Hammer and Fluid condition: A computational approach, 1st Festival on Water and Wastewater Research and Technology, Tehran, Iran, 12–17 Dec. **2009**, 27–34, http://isrc.nww.co.ir.

72. Hariri Asli, K.; Nagiyev, F. B.; Haghi, A. K.; Aliyev, S. A.; A computational Method to Study Transient Flow in Binary Mixtures, 1st Festival on Water and Wastewater Research and Technology, Tehran, Iran, 12–17 Dec. **2009,** 27–35, http://isrc.nww.co.ir.

73. Hariri Asli, K.; Nagiyev, F. B.; Haghi, A. K.; Physical modeling of fluid movement in pipelines, 1st Festival on Water and Wastewater Research and Technology, Tehran, Iran, 12–17 Dec. **2009**, 27–36, http://isrc.nww.co.ir.

74. Hariri Asli, K.; Nagiyev, F. B.; Haghi, A. K.; Aliyev, S. A.; Interpenetration of two fluids at parallel between plates and turbulent moving, 1st Festival on Water and Wastewater Research and Technology, Tehran, Iran, 12–17 Dec. **2009**, 27–37, http://isrc.nww.co.ir.

75. Hariri Asli, K.; Nagiyev, F. B.; Haghi, A. K.; Aliyev, S. A.; Modeling of fluid interaction produced by water hammer, 1st Festival on Water and Wastewater Research and Technology, Tehran, Iran, 12–17 Dec. **2009,** 27–38, http://isrc.nww.co.ir.

76. Hariri Asli, K.; Nagiyev, F. B.; Haghi, A. K.; Aliyev, S. A.; GIS and water hammer disaster at earthquake in Rasht pipeline, 1st Festival on Water and Wastewater Research and Technology, Tehran, Iran, 12–17 Dec. **2009,** 27–39, http://isrc.nww.co.ir.

77. Hariri Asli, K.; Nagiyev, F. B.; Haghi, A. K.; Aliyev, S. A.; Interpenetration of two fluids at parallel between plates and turbulent moving, 1st Festival on Water and Wastewater Research and Technology, Tehran, Iran, 12–17 Dec.2009, 27–40, http://isrc.nww.co.ir.

78. Hariri Asli, K.; Nagiyev, F. B.; Haghi, A. K.; Aliyev, S. A.; Water hammer and hydrodynamics' instability, 1st Festival on Water and Wastewater Research and Technology, Tehran, Iran, 12–17 Dec.2009, 27–41, http://isrc.nww.co.ir.

79. Hariri Asli, K.; Nagiyev, F. B.; Haghi, A. K.; Aliyev, S. A.; Water hammer analysis and formulation, 1st Festival on Water and Wastewater Research and Technology, Tehran, Iran, 12–17 Dec. **2009,** 27–42, http://isrc.nww.co.ir.

80. Hariri Asli, K.; Nagiyev, F. B.; Haghi, A. K.; Aliyev, S. A.; Water hammer and fluid condition, 1st Festival on Water and Wastewater Research and Technology, Tehran, Iran, 12–17 Dec.2009, 27–43, http://isrc.nww.co.ir.

81. Hariri Asli, K.; Nagiyev, F. B.; Haghi, A. K.; Aliyev, S. A.; Water hammer and pump pulsation, 1st Festival on Water and Wastewater Research and Technology, Tehran, Iran, 12–17 Dec. **2009,** 27–44, http://isrc.nww.co.ir.

82. Hariri Asli, K.; Nagiyev, F. B.; Haghi, A. K.; Aliyev, S. A.; Reynolds number and hydrodynamics instabilities, 1st Festival on Water and Wastewater Research and Technology, Tehran, Iran, 12–17 Dec. **2009,** 27–45, http://isrc.nww.co.ir.

83. Hariri Asli, K.; Nagiyev, F. B.; Haghi, A. K.; Aliyev, S. A.; water hammer and valves, 1st Festival on Water and Wastewater Research and Technology, Tehran, Iran, 12–17 Dec. **2009,** 27–46, http://isrc.nww.co.ir.

84. Hariri Asli, K.; Nagiyev, F. B.; Haghi, A. K.; Aliyev, S. A.; "Water hammer and fluid Interpenetration," 1st Festival on Water and Wastewater Research and Technology, Tehran, Iran, 12–17 Dec. **2009,** 27–47, http://isrc.nww.co.ir.

85. Hariri Asli, K.; Nagiyev, F. B.; Modeling of fluid interaction produced by water hammer, International Journal of Chemoinformatics and Chemical Engineering, IGI, ISSN: 2155–4110, EISSN: 2155–4129, USA, **2010,** 29–41, http://www.igi-global.com/journals/details.asp?ID=34654

86. Hariri Asli, K.; Nagiyev, F. B.; Haghi, A. K.; Water hammer and fluid condition; a computational approach, Computational Methods in Applied Science and Engineering, USA, Chapter 5, Nova Science Publications, ISBN: 978-1-60876-052-7, USA, **2010,** 73–94, https://www.novapublishers.com/catalog/

87. Hariri Asli, K.; Nagiyev, F. B.; Haghi, A. K.; Some aspects of physical and numerical modeling of water hammer in pipelines. Computational Methods in Applied Science and Engineering, USA, Chapter 23, Nova Science Publications, ISBN: 978-1-60876-052-7, USA, **2010,** 365–387, https://www.novapublishers.com/catalog/

88. Hariri Asli, K.; Nagiyev, F. B.; Haghi, A. K.; Modeling for water hammer due to valves; from theory to practice, Computational Methods in Applied Science and En-

gineering, USA, Chapter 11, Nova Science Publications ISBN: 978-1-60876-052-7, USA, **2010**, 229–236, https://www.novapublishers.com/catalog/

89. Hariri Asli, K.; Nagiyev, F. B.; Haghi, A. K.; A computational method to Study transient flow in binary mixtures, Computational Methods in Applied Science and Engineering, USA, Chapter 13, Nova Science Publications ISBN: 978-1-60876-052-7, USA, **2010**, 229–236, https://www.novapublishers.com/catalog/

90. Hariri Asli, K.; Nagiyev, F. B.; Haghi, A. K.; Water hammer analysis; some computational aspects and practical hints, Computational Methods in Applied Science and Engineering, USA, Chapter 16, Nova Science Publications ISBN: 978-1-60876-052-7, USA, **2010**, 263–281, https://www.novapublishers.com/catalog/

91. Hariri Asli, K.; Nagiyev, F. B.; Haghi, A. K.; Water hammer and hydrodynamics instabilities modeling, Computational Methods in Applied Science and Engineering, USA, Chapter 17, From Theory to Practice, Nova Science Publications ISBN: 978-1-60876-052-7, USA, **2010**, 283–301, https://www.novapublishers.com/catalog/

92. Hariri Asli, K.; Nagiyev, F. B.; Haghi, A. K.; A computational approach to study water hammer and pump pulsation phenomena, Computational Methods in Applied Science and Engineering, USA, Chapter 22, Nova Science Publications, ISBN: 978-1-60876-052-7, USA, **2010**, 349–363, https://www.novapublishers.com/catalog/

93. Hariri Asli, K.; Nagiyev, F. B.; Haghi, A. K.; A computational approach to study fluid movement, Nanomaterials Yearbook – **2009**, From Nanostructures, Nanomaterials and Nanotechnologies to Nanoindustry, Chapter 16, Nova Science Publications, USA, ISBN: 978-1-60876-451-8, USA, **2010**, 181–196, https://www.novapublishers.com/catalog/product_info.php?products_id=11587

94. Hariri Asli, K.; Nagiyev, F. B.; Haghi, A. K.; Physical modeling of fluid movement in pipelines, Nanomaterials Yearbook – **2009**, From Nanostructures, Nanomaterials and Nanotechnologies to Nanoindustry, Chapter 17, Nova Science Publications, USA, ISBN: 978-1-60876-451-8, USA, **2010**, 197–214, https://www.novapublishers.com/catalog/product_info.php?products_id=11587

95. Hariri Asli, K.; Nagiyev, F. B.; Haghi, A. K.; "Some Aspects of Physical and Numerical Modeling of water hammer in pipelines," Nonlinear Dynamics An International Journal of Nonlinear Dynamics and Chaos in Engineering Systems, ISSN: 1573–269X (electronic version) Journal no. 11071 Springer, Netherlands, **2009**, ISSN: 0924–090X (print version), Springer, Heidelberg, Germany, Number 4 / June, **2010**, 60, 677–701, http://www.springerlink.com/openurl.aspgenre=article&id=doi: 10.1007/s11071-009-9624-7.

96. Hariri Asli, K.; Nagiyev, F. B.; Haghi, A. K.; Interpenetration of two fluids at parallel between plates and turbulent moving in pipe; a case study, Computational Methods in Applied Science and Engineering, USA, Chapter 7, Nova Science Publications, ISBN: 978-1-60876-052-7, USA, **2010**, 107–133, https://www.novapublishers.com/catalog/

97. Hariri Asli, K.; Nagiyev, F. B.; Beglou, M. J.; Haghi, A. K.; Kinetic analysis of convective drying, International Journal of the Balkan Tribological Association, ISSN: 1310–4772, Sofia, Bulgaria, **2009**, 15, 4, 546–556, jbalkta@gmail.com

98. Hariri Asli, K.; Nagiyev, F. B.; Haghi, A. K.; Three-dimensional Conjugate Heat Transfer in Porous Media, International Journal of the Balkan Tribological Association, ISSN: 1310–4772, Sofia, Bulgaria, **2009**, 15(3), 336–346, jbalkta@gmail.com

99. Hariri Asli, K.; Nagiyev, F. B.; Haghi, A. K.; Aliyev, S. A.; Pure Oxygen penetration in wastewater flow, Recent Progress in Research in Chemistry and Chemical Engineering, Nova Science Publications, ISBN: 978-1-61668-501-0, Nova Science Publications, USA, **2010**, 17–27, https://www.novapublishers.com/catalog/product_info. php?products_id=13174110100. Hariri Asli, K.; Nagiyev, F. B.; Haghi, A. K.; Aliyev, S. A.; Improved modeling for prediction of water transmission failure, Recent Progress in Research in Chemistry and Chemical Engineering, Nova Science Publications, ISBN: 978-1-61668-501-0, Nova Science Publications, USA, **2010**, 28–36, https:// www.novapublishers.com/catalog/product_info.php?products_id=13174

101. Harms, T. M.; Kazmierczak, M. J.; Cerner, F. M.; Holke A.; Henderson, H. T.; Pilchowski, H. T.; Baker K.; Experimental Investigation of Heat Transfer and Pressure Drop through Deep Micro channels in a (100) Silicon Substrate, in: Proceedings of the ASME.; Heat Transfer Division, HTD **1997,** *351,* 347–357.

102. Holland, F. A.; Bragg R.; Fluid Flow for Chemical Engineers, Edward Arnold Publishers, London, **1995,** 1–3.

103. George E.; Glimm J.; "Self-similarity of Rayleigh-Taylor mixing rates,"*Phys. Fluids* **2005,** *17,* 054101, 1–3.

104. Goncharov V. N.; "Analytical model of nonlinear, single-mode, classical Rayleigh-Taylor instability at arbitrary Atwood numbers,"*Phys. Rev. Lett.* 88, 134502, **2002,** 10–15.

105. Ishii M.; Thermo-Fluid Dynamic Theory of Two-Phase Flow, Collection de D. R.; Liles Reed, W. H.; "A Sern-Implict Method for Two-Phase Fluid la Direction des Etudes et. Recherché d'Electricite de France, 22 Dynamics,"*J. Computational Physics Paris,* **1975,** *26,* 390–407.

106. Jaime Suárez A.; "Generalized water hammer algorithm for piping systems with unsteady friction" **2005,** 72–77.

107. Jaeger C.; "Fluid Transients in Hydro-Electric Engineering Practice," Blackie and Son Ltd.; **1977,** 87–88.

108. Joukowski N.; Paper to Polytechnic Soc. Moscow, spring of1898, English translation by Miss O.; Simin. Proc. AWWA, **1904,** 57–58.

109. Kodura A.; Weinerowska K.; the influence of the local pipeline leak on water hammer properties, Materials of the II Polish Congress of Environmental Engineering, Lublin, **2005,** 125–133.

110. Kane J.; Arnett D.; Remington, B. A.; Glendinning, S. G.; Baz'an G.; "Two-dimensional versus three-dimensional supernova hydrodynamic instability growth," Astrophys. J.; **2000,** 528–989.

111. Karassik, I. J.; "Pump Handbook – Third Edition," McGraw-Hill, **2001,** 19–22.

112. Koelle E.; Luvizotto Jr. E.; Andrade, J. P. G.; "Personality Investigation of Hydraulic Networks using MOC – Method of Characteristics" Proceedings of the 7th International Conference on Pressure Surges and Fluid Transients, Harrogate Durham, United Kingdom, **1996,** 1–8.

113. Kerr, S. L.; "Minimizing service interruptions due to transmission line failures: Discussion," Journal of the American Water Works Association, *41, 634,* July **1949,** 266–268.

114. Kerr, S. L.; "Water hammer control," Journal of the American Water Works Association, *43,* December **1951,** 985–999.

115. Kraichnan, R. H.; Montgomery D.; "Two-dimensional turbulence," Rep. Prog. Phys.1967, *43, 547*, 1417–1423.

116. Leon Arturo, S.; "An efficient second-order accurate shock-capturing scheme for modeling one and two-phase water hammer flows" PhD Thesis, March29, **2007**, 4–44.

117. Lee, T. S.; Pejovic, S. **1996,** Air influence on similarity of hydraulic transients and vibrations. *ASME J. Fluid Eng.* 118(4), 706–709.

118. Li J.; McCorquodale A.; "Modeling Mixed Flow in Storm Sewers," Journal of Hydraulic Engineering, ASCE, *125,* No. *11,* **1999,** 1170–1180.

119. Mala, G.; Li, D.; Dale, J. D.; Heat Transfer and Fluid Flow in Microchannels, J.; Heat Transfer, *40,* **1997,** 3079–3088.

120. Miller, P. L.; Cabot, W. H.; Cook, A. W.; "Which way is up? A fluid dynamics riddle,"*Phys. Fluids 17,* 091110, **2005,** 1–26.

121. Minnaert, M.; on musical air bubbles and the sounds of running water. Phil. Mag.; **1933,** *16(7),* 235–248.

122. Moeng, C. H.; McWilliams, J. C.; Rotunno, R.; Sullivan, P. P.; Weil, J.; "Investigating 2D modeling of atmospheric convection in the PBL,"*J. Atm. Sci. 61,* **2004,** 889 −903.

123. Moody, L. F.; "Friction Factors for Pipe Flow," Trans. ASME, **1944,** *66,* 671–684.

124. Nagiyev, F. B.; Dynamics, heat and mass transfer of vapor-gas bubbles in a two-component liquid. Turkey-Azerbaijan petrol semin.; Ankara, Turkey, **1993,** 32–40.

125. Nagiyev, F. B.; The method of creation effective coolness liquids, Third Baku international Congress. Baku, Azerbaijan Republic, **1995,** 19–22.

126. Neshan H.; Water Hammer, pumps Iran Co. Tehran, Iran, **1985,** 1–60.

127. Nigmatulin, R. I.; Khabeev, N. S.; Nagiyev, F. B.; Dynamics, heat and mass transfer of vapor-gas bubbles in a liquid. Int. J.; Heat Mass Transfer, vol.24, N6, Printed in Great Britain, **1981,** 1033–1044.

128. Oron, D.; Arazi, L.; Kartoon, D.; Rikanati, A.; Alon, U.; Shvarts, D.; "Dimensionality dependence of the Rayleigh-Taylor and Richtmyer-Meshkov instability late-time scaling laws,"*Phys. Plasmas* **2001,** *8,* 2883p.

129. Parmakian, J.; "Water hammer Design Criteria,"*J. Power Div.ASCE,* Sept.; **1957,** 456–460.

130. Parmakian, J.; "Water hammer Analysis," Dover Publications, Inc.; New York, **1963,** 51–58.

131. Perry.; R. H.; Green, D. W.; Maloney, J. O.; Perry's Chemical Engineers Handbook, 7th Edition, McGraw-Hill, New York, **1997,** 1–61.

132. Peng, X. F.; Peterson, G. P.; Convective Heat Transfer and Flow Friction for Water Flow in Microchannel Structure, Int. J.; Heat Mass Transfer *36,* **1996,** 2599–2608.

133. Pickford J.; "Analysis of Surge," Macmillian, London, **1969,** 153–156.

134. Pipeline Design for Water and Wastewater, American Society of Civil Engineers, New York, **1975,** 54 p.

135. Qu W.; Mala, G. M.; Li D.; Heat Transfer for Water Flow in Trapezoidal Silicon Microchannels, **1993,** 399–404.

136. Quick, R. S.; "Comparison and Limitations of Various Water hammer Theories,"*J. Hyd. Div. ASME,* May, **1933,** 43–45.

137. Rayleigh, On the pressure developed in a liquid during the collapse of a spherical cavity. Philos. Mag. Ser.6, **1917,** *34(200),* 94–98.

138. Ramaprabhu P.; Andrews, M. J.; "Experimental investigation of Rayleigh-Taylor mixing at small Atwood numbers,"*J. Fluid Mech.***2004,** *502,* 233 p.
139. Rich, G. R.; "Hydraulic Transients," Dover, USA, **1963,** 148–154.
140. Savic, D. A.; Walters, G. A.; "Genetic Algorithms Techniques for Calibrating Network Models," Report No. 95/12, Centre for Systems and Control Engineering, **1995,** 137–146.
141. Savic, D. A.; Walters, G. A.; Genetic Algorithms Techniques for Calibrating Network Models, University of Exeter, Exeter, United Kingdom, **1995,** 41–77.
142. Shvarts D.; Oron D.; Kartoon D.; Rikanati A.; Sadot O.; "Scaling laws of non-linear Rayleigh-Taylor and Richtmyer-Meshkov instabilities in two and three dimensions,"C. R. Acad. Sci. Paris, IV, **2000,** *719,* 312 p.
143. Sharp, B.; "Water hammer Problems and Solutions," Edward Arnold Ltd.; London, **1981,** 43–55.
144. Skousen, P.; "Valve Handbook," McGraw Hill, New York, HAMMER Theory and Practice, **1998,** 687–721.
145. Song, C. C. et al.; "Transient Mixed-Flow Models for Storm Sewers,"*J.* Hyd. Div.; Nov.; **1983,** *109,* 458–530.
146. Stephenson D.; "Pipe Flow Analysis," Elsevier, *19,* S.A.; **1984,** 670–788.
147. Streeter, V. L.; Lai C.; "Water hammer Analysis Including Fluid Friction." *J. Hydraulics Division, ASCE, 88,* **1962,** 79 p.
148. Streeter, V. L.; Wylie, E. B.; "Fluid Mechanics," McGraw-Hill Ltd.; USA, **1979,** 492–505.
149. Streeter, V. L.; Wylie, E. B.; "Fluid Mechanics," McGraw-Hill Ltd.; USA, **1981,** 398–420.
150. Tijsseling,"Alan E Vardy Time scales and FSI in unsteady liquid-filled pipe flow," **1993,** 5–12.
151. Tuckerman, D. B.; R. F. W Pease, high performance heat sinking for VLSI, IEEE Electron device letter, DEL-2, **1981,** 126–129.
152. Tennekes H.; Lumley, J. L.; A First Course in Turbulence, the MIT Press, **1972,** 410–466.
153. Thorley, A. R. D.; "Fluid Transients in Pipeline Systems," D. & L. George, Herts, England, **1991,** 231–242.
154. Tullis, J. P.; "Control of Flow in Closed Conduits," Fort Collins, Colorado, **1971,** 315–340.
155. Tuckerman, D. B.; Heat transfer microstructures for integrated circuits, Ph.D. thesis, Stanford University, 1984, 10–120.
156. Vallentine, H. R.; "Rigid Water Column Theory for Uniform Gate Closure,"J. Hyd. Div. ASCE, July, 1965, 55–243.
157. Waddell, J. T.; Niederhaus, C. E.; Jacobs, J. W.; "Experimental study of Rayleigh-Taylor instability: Low Atwood number liquid systems with single-mode initial perturbations,"Phys. Fluids 13, 2001, 1263–1273.
158. Watters, G. Z.; "Modern Analysis and Control of Unsteady Flow in Pipelines,"Ann. Arbor Sci.; 2nd Ed.; 1984, 1098–1104.
159. Walski, T. M.; Lutes, T. L.; "Hydraulic Transients Cause Low-Pressure Problems," Journal of the American Water Works Association, 75(2), 1994, 58.

160. Weber, S. V.; Dimonte G.; Marinak, M. M.; "Arbitrary Lagrange-Eulerian code simulations of turbulent Rayleigh-Taylor instability in two and three dimensions," Laser and Particle Beams 21, 2003, 455 p.

161. Wood, D. J.; Dorsch, R. G.; Lightener C.; "Wave-Plan Analysis of Unsteady Flow in Closed Conduits," Journal of Hydraulics Division, ASCE, 92, 1966, 83–110.

162. Wood, D. J.; Jones, S. E. "Water hammer charts for various types of valves," Journal of the Hydraulics Division, Proceedings of the American Society of Civil Engineers, January, 1973, 167–178.

163. Wood, F. M.; "History of Water hammer," Civil Engineering Research Report, #65, Queens University, Canada, 1970, 66–70.

164. Wu, Z. Y.; Simpson, A. R.; Competent genetic-evolutionary optimization of water distribution systems. J. Comput. Civ. Eng.2001, 15(2), 89–101.

165. Wylie, E. B.; Talpin, L. B.; Matched impedance to control fluid transients. Trans. ASME1983, 105(2), 219–224.

166. Wylie, E. B.; Streeter, V. L.; Fluid Transients in Systems, Prentice-Hall, Englewood Cliffs, New Jersey, 1993, 4 p.

167. Wylie, E. B.; Streeter, V. L.; Fluid Transients, Feb Press, Ann. Arbor, MI, 1982, corrected copy, 1983, 158 p.

168. Wu, P. Y.; Little, W. A.; measurement of friction factor for flow of gases in very fine channels used for micro miniature, Joule Thompson refrigerators, Cryogenics 1983, 24(8), 273–277.

169. Xu B.; Ooi, K. T.; Mavriplis, C.; Zaghloul, M. E.; Viscous dissipation effects for liquid flow in microchannels, Micorsystems, 2002, 53–57.

170. Young, Y. N.; Tufo, H.; Dubey, A.; Rosner, R.; "On the miscible Rayleigh-Taylor instability: two and three dimensions,"J. Fluid Mech.2001, 447, 377, 2003–2500.

CHAPTER 4

HEAT TRANSFER AND VAPOR BUBBLE

CONTENTS

4.1   INTRODUCTION

In the two phases flow is extremely important to the concept of volume concentration. This is the relative volume fraction of one phase in the volume of the pipe [1]. Such an environment typical fluid is a high density and little compressibility. This property contributes to the creation of various forms of transient conditions [2-3]. Figure 1 shows an experimental setup, which investigated the formation of different modes of fluid flow with gas bubbles and steam.

4.2   MATERIALS AND METHODS

The experimental results show that the bubble flow usually occurs at low concentrations of vapor. It includes three main types of flow regimes in microgravity bubble, slug and an annular.

TABLE 1   Simulators, models and problems.

Cases	Range of problems
Steady or gradually varying turbulent flow	Rapidly varying or transient flow
Incompressible, Newtonian, single-phase fluids	Slightly compressible, two-phase fluids (vapor and liquid) and two-fluid systems (air and liquid)
Full pipes	Closed-conduit pressurized systems with air intake and release at discrete points

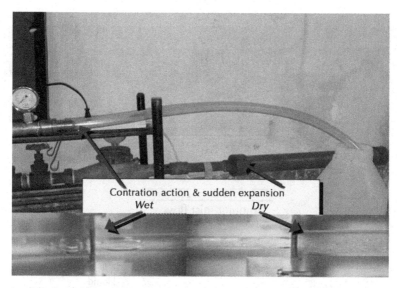

FIGURE 1 Snapshot laboratory setup for studying the structure of flow in different configurations tube.

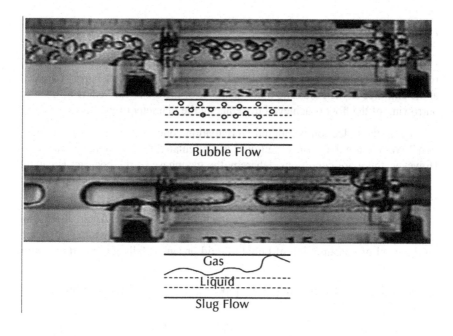

FIGURE 2 *(Continued)*

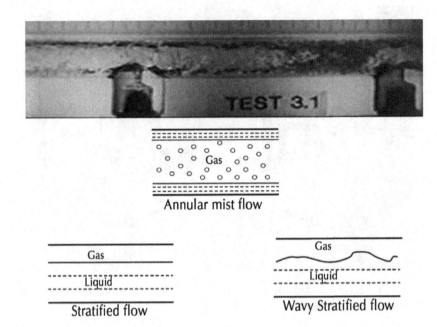

FIGURE 2   Types of flows.

Figure 2 shows snapshots of the flow pattern for various configurations of the tube [4, 5]. The flow enters the tee at the bottom of the picture, and then is divided at the entrance to the tube. The inner diameter of the tee is 1.27 cm (Fig. 1). The narrowing of the flow is achieved by reducing the diameter of the hose.

Within the reduction is a liquid recirculation zone, called the "vena contraction." Wet picture that, when the liquid is recirculated to the "vena contraction." However, there are conditions, whereby the gas phase of the contract is caught vena. Figure 3 shows the different flow regimes observed in the experimental setup [6].

Increased flow is achieved by increasing the size of the pipe. Again, there is an area of the liquid recirculation near the "corner" a sudden expansion. Depending on the level of consumption of bubbly liquid or gas, it falls into the trap in this area.

In the inlet fluid moves out of the pipe diameter of 12.7 (*mm*) in the pipe diameter of 25 (*mm*).

Normal extension occurs at the beginning of the flow. Soon comes the expansion section, and the flow rate continues to increase. Two-phase jet stream created, ultimately, with areas of air flowing above and below the bubble region [6]. The behavior of gas-liquid mixture in the expansion is proportional to increasing the diameter of the pipe [7, 8].

It is shown in Figs. 1-3 that in place of the sudden expansion of a transition flow regimes of the turbulent flow. Depending on the flow or gas bubble mixture it falls into the trap in this area.

Experiments were conducted on the pipe, whose diameter is suddenly doubled [9]. In this case the region of turbulence of fluid are observed around the "corner" a sudden expansion. The expansion is observed at the beginning of the flow. As a result of turbulence flow gap expansion increases and the flow rate continues to increase. In the end, creates a stream of two-phase flow, with airfields, the current above and below the bubble area [10, 11].

Normal

Gap

Jet

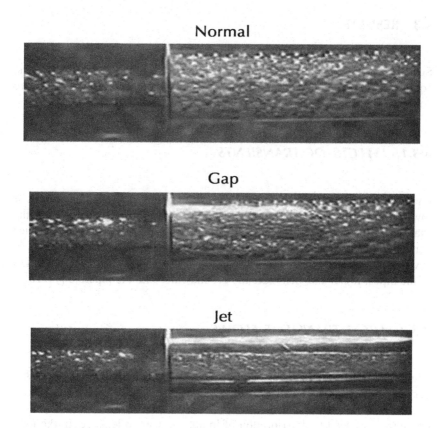

FIGURE 3   Narrowing and sudden expansion flow level.

With this experimental setup is shown that the formation of different modes of two-phase flow depends on the relative concentration of these phases and the flow rate. Figures 1 and 2 shows a diagram of core flow of vapor-liquid flow regimes, in particular, the bubble, stratified, slug, stratified, and the wave dispersion circular flow. In these experiments the mode of vapor-liquid flow in a pipe, when the bubbles are connected in long steam field, whose dimensions are commensurate with the diameter of the pipe [12-13].This flow is called the flow of air from the tube. In the transition from moderate to high-speed flow, when the concentration of vapor and liquid are approximately equal, the flow regime is often irregular and even chaotic[14, 15]. By the simulated conditions, It is assumed that the electricity suddenly power off without warning (i.e., no time to turn the diesel generators or pumps) [15, 16].

4.3   RESULTS

Such situations are the strong reason of the installation of pressure sensors, equipped with high-speed data loggers. Therefore, the following items are consequences, which may result in these situations.

4.3.1   EFFECTS OF TRANSIENTS

Hydraulic transients can lead to the following physical phenomena. High or low transient pressures that may arise in the piping and connections in the share of second. They often alternate from highest to lowest levels and vice versa. High pressures are a consequence of the collapse of steam bubbles or cavities is similar to steam pump cavitations. It can yield the tensile strength of the pipes. It can also penetrate the groundwater into the pipeline [17-18].

4.3.2   HIGH TRANSIENT FLOWS

High-speed flows are also very fast pulse pressure. It leads to temporary but very significant transient forces in the bends and other devices that can make a connection to deform. Even strain buried pipes under the influence of cyclical pressures may lead to deterioration of joints and lead to leakage. In the low-pressure pumping stations at downstream a very rapid closing of the valve, known as shut off valve, may lead to high pressure transient flows.

4.3.3 WATER COLUMN SEPARATION

Water column, usually are separated with sharp changes in the profile or the local high points. It is because of the excess of atmospheric pressure. The spaces between the columns are filled with water or the formation of steam (e.g., steam at ambient temperature) or air, if allowed admission into the pipe through the valve. Collapse of cavitation bubbles or steam can cause the dramatic impact of rising pressure on the transition process.

If the water column is divided very quickly, it could in turn lead to rupture of the pipeline. Vapors cavitation may also lead to curvature of the pipe. High-pressure wave can also be caused by the rapid removal of air from the pipeline.

Steam bubbles or cavities are generated during the hydraulic transition. The level of hydraulic pressure (EGD) or pressure in some areas could fall low enough to reach the top of the pipe. It leads to subatmospheric pressure or even full-vacuum pressures. Part of the water may undergo a phase transition, changing from liquid to steam, while maintaining the vacuum pressure.

This leads to a temporary separation of the water column. When the system pressure increases, the columns of water rapidly approach to each other. The pair reverts to the liquid until vapor cavity completely dissolved. This is the most powerful and destructive power of water hammer phenomenon.

4.3.4 GLOBAL REGULATION OF STEAM PRESSURE

If system pressure drops to vapor pressure of the liquid, the fluid passes into the vapor, leading to the separation of liquid columns. Consequently, the vapor pressure is a fundamental parameter for hydraulic transient modeling. The vapor pressure varies considerably at high temperature or altitude. Fortunately, for typical water pipelines and networks, the pressure does not reach such values. If the system is at high altitude or if it is the industrial system, operating at high temperatures or pressures, it should be guided by a table or a state of vapor pressure curve vapor-liquid.

4.3.5 VIBRATION

Pressure fluctuations associated with the peculiarities of the system, as well as the peculiarities of its design. The pump must be assumed just as one of the promoters of the system. Effect of pressure relief valve on the fluctuations of the liquid often turned out three times more damaging than the effect of the pump. Such monitoring fluid flow controlling means usually has more negative impact than the influence of the pump.

Rapid changes in the transition pressures can lead to fluctuations or resonance. It can damage the pipe, resulting in leakage or rupture. Experiments show that the flows in the pipe will be very small, say, located at 24 km/h.

This corresponds to about 0.45% of the velocity of pressure. In this case, the flow fluctuations can be easily accumulated and redeemed until the next perturbation. Fluctuations of the flow are no fluctuations of pressure.

In the case when the source of the pressure fluctuations is the so-called, "Acceleration factor," one can say that in order to accelerate the mass of fluid in the system until the new rate of additional efforts.

4.3.6 BASIC EQUATIONS DESCRIBING THE SPHERICALLY SYMMETRIC MOTION OF A BUBBLE BINARY SOLUTION

The dynamics and heat and mass transfer of vapor bubble in a binary solution of liquids, in Ref. [8], was studied for significant thermal, diffusion and inertial effect.

It was assumed that binary mixture with a density ρ_1, consisting of components 1 and 2, respectively, the density ρ_1 and ρ_2.

Moreover,

$$\rho_1 + \rho_2 = \rho_1,$$

where, the mass concentration of component 1 of the mixture.

Also [8] consider a two-temperature model of interphase heat exchange for the bubble liquid. This model assumes homogeneity of the temperature in phases.

The intensity of heat transfer for one of the dispersed particles with an endless stream of carrier phase will be set by the dimensionless parameter of Nusselt Nu_1.

Bubble dynamics described by the Rayleigh equation:

$$R\dot{w}_l + \frac{3}{2}w_l^2 = \frac{p_1 + p_2 - p_\infty - 2\sigma/R}{\rho_l} - 4v_1\frac{w_l}{R} \qquad (1)$$

where ρ_1 and ρ_2 – the pressure component of vapor in the bubble,

p∞ – the pressure of the liquid away from the bubble,

σ and v_1 – surface tension coefficient of kinematic viscosity for the liquid.

Consider the condition of mass conservation at the interface.

Mass flow j_i^{TH} component $(i = 1,2)$ of the interface $r = R(t)$ in j_i^{TH} phase per unit area and per unit of time and characterizes the intensity of the phase transition is given by:

$$j_i = \rho_i \left(\dot{R} - w_l - w_i \right), \; (i = 1,2),$$ (2)

where,

w_i– The diffusion velocity component on the surface of the bubble. The relative motion of the components of the solution near the interface is determined by Fick's law:

$$\rho_1 w_1 = -\rho_2 w_2 = -\rho_l D \frac{\partial k}{\partial r}\bigg|_R$$ (3)

If we add Eq. (2), while considering that:

$\rho_1 + \rho_2 = \rho_1$, and draw the Eq. (3), we obtain

$$\dot{R} = w_l + \frac{j_1 + j_2}{\rho_l},$$ (4)

Multiplying the first Eq. (2) on ρ_2, the second in ρ_2 and subtract the second equation from the first. In view of Eq. (3) we obtain:

$$k_R j_2 - (1 - k_R) j_1 = -\rho_l D \frac{\partial k}{\partial r}\bigg|_R$$

where k_R – the concentration of the first component at the interface. With the assumption of homogeneity of parameters inside the bubble changes in the mass of each component due to phase transformations can be written as

$$\frac{d}{dt}\left(\frac{4}{3}\pi R^3 \rho_i'\right) = 4\pi R^2 j_i, \text{ or}$$

$$\frac{R}{3}\dot{\rho}_i' + \dot{R}\rho_i' = j_i, \ (i = 1,2),\tag{5}$$

Express the composition of a binary mixture in mole fractions of the component relative to the total amount of substance in liquid phase

$$N = \frac{n_1}{n_1 + n_2},\tag{6}$$

The number of moles i^{TH} component n_i, which occupies the volume V, expressed in terms of its density:

$$n_i = \frac{\rho_i V}{\mu_i},\tag{7}$$

Substituting Eq. (7) in Eq. (6), we obtain

$$N_1(k) = \frac{\mu_2 k}{\mu_2 k + \mu_1(1-k)},\tag{8}$$

By law, Raul partial pressure of the component above the solution is proportional to its molar fraction in the liquid phase, i.e.,

$$p_1 = p_{S1}(T_v)N_1(k_R), \ p_2 = p_{S2}(T_v)[1 - N_1(k_R)],\tag{9}$$

Equations of state phases have the form:

$$p_i = BT_v \rho_i' / \mu_i, \ (i = 1,2),\tag{10}$$

where,

B – Gas constant,

T_v – The temperature of steam,

ρ_i' – The density of the mixture components in the vapor bubble,

μ_i – Molecular weight,

p_{si} – Saturation pressure.

The boundary conditions r = ∞ and on a moving boundary can be written as

$$k\big|_{r=\infty} = k_0, \; k\big|_{r=R} = k_R, \; T_l\big|_{r=\infty} = T_0, \; T_l\big|_{r=R} = T_v, \tag{11}$$

$$j_1 l_1 + j_2 l_2 = \lambda_l D \frac{\partial T_l}{\partial r}\bigg|_{r=R}, \tag{12}$$

where, l_i – specific heat of vaporization.

By the definition of Nusselt parameter – the dimensionless parameter characterizing the ratio of particle size and the thickness of thermal boundary layer in the phase around the phase boundary and determined from additional considerations or from experience.

The heat of the bubble's intensity with the flow of the carrier phase will be further specified as:

$$\left(\lambda_l \frac{\partial T_l}{\partial r}\right)_{r=R} = Nu_l \cdot \frac{\lambda_l (T_0 - T_v)}{2R}, \tag{13}$$

In [16] obtained an analytical expression for the Nusselt parameter:

$$Nu_l = 2\sqrt{\frac{\omega R_0^2}{a_l}} = 2\sqrt{\frac{R_0}{a_l}}\sqrt{\frac{3\gamma p_0}{\rho_l}} = 2\sqrt{\sqrt{3\gamma} \cdot Pe_l}, \tag{14}$$

where, $a_l = \dfrac{\lambda_l}{\rho_l c_l}$ – thermal diffusivity of fluid,

$$Pe_l = \frac{R_0}{a_l}\sqrt{\frac{p_0}{\rho_l}} - \text{Peclet number.}$$

The intensity of mass transfer of the bubble with the flow of the carrier phase will continue to ask by using the dimensionless parameter Sherwood Sh:

$$\left(D\frac{\partial k}{\partial r}\right)_{r=R} = Sh \cdot \frac{D(k_0 - k_R)}{2R}$$

where, D – diffusion coefficient,

k – The concentration of dissolved gas in liquid,

The subscripts 0 and R refer to the parameters in an undisturbed state and at the interface.

We define a parameter in the form of Sherwood [16]

$$Sh = 2\sqrt{\frac{\omega R_0^2}{D}} = 2\sqrt{\frac{R_0}{D}}\sqrt{\frac{3\gamma p_0}{\rho_l}} = 2\sqrt{\sqrt{3\gamma} \cdot Pe_D}, \qquad (15)$$

where $Pe_D = \frac{R_0}{D}\sqrt{\frac{p_0}{\rho_l}}$ – diffusion Peclet number.

The system of Eqs. (1)-(15) is a closed system of equations describing the dynamics and heat transfer of insoluble gas bubbles with liquid[19-58].

4.3.7 THE BRAKING EFFECT OF THE INTENSITY OF PHASE TRANSFORMATIONS IN BOILING BINARY SOLUTIONS

If we use Eqs.(7)–(9), we obtain relations for the initial concentration of component 1:

$$k_0 = \frac{1 - \chi_2^0}{1 - \chi_2^0 + \mu(\chi_1^0 - 1)}, \ \mu = \mu_2 / \mu_1, \ \chi_i^0 = p_{si0} / p_0, \ i = 1,2, \quad (16)$$

where,

μ_2, μ_1 – Molecular weight of the liquid components of the mixture, P_{si0} – saturated vapor pressure of the components of the mixture at an initial temperature of the mixture T_0, which were determined by integrating the Clausius-Clapeyron relation. The parameter χ_i^0 is equal to

$$\chi_i^0 = \exp\left[\frac{l_i \mu_i}{B}\left(\frac{1}{T_{ki}} - \frac{1}{T_0}\right)\right], \quad (17)$$

Gas-phase liquid components in the derivation of Eq. (2) seemed perfect gas equations of state:

$$p_i = \rho_i B T_i / \mu_i.$$

where B – universal gas constant,

P_i– the vapor pressure inside the bubble T_i to the temperature in the ratio of Eq. (2),

T_{ki} – temperature evaporating the liquid components of binary solution at an initial pressure p_0,

l_i– specific heat of vaporization.

The initial concentration of the vapor pressure of component p_0 is determined from the relation:

$$c_0 = \frac{k_0 \chi_1^0}{k_0 \chi_1^0 + (1 - k_0)\chi_2^0}, \quad (18)$$

In this paper the problem of radial motions of a vapor bubble in binary solution was solved. It was investigated at various pressure drops in the liquid for different initial radii R_0 for a bubble. It is of great practical interest of aqueous solutions of ethanol and ethylene glycol.

It was revealed an interesting effect. The parameters characterized the dynamics of bubbles in aqueous ethyl alcohol. It was studied in the field of variable pressure lie between the limiting values of the parameter p_0 for pure components [59-68].

The pressure drops and consequently the role of diffusion are assumed unimportant. The pressure drop along with the heat dissipation is included diffusion dissipation. The rate of growth and collapse of the bubble is much higher than in the corresponding pure components of the solution under the same conditions. A completely different situation existed during the growth and collapse of vapor bubble in aqueous solutions of ethylene glycol.

In this case, the effect of diffusion resistance, leaded to inhibition of the rate of phase transformations. The growth rate and the collapse of the bubble is much smaller than the corresponding values, but for the pure components of the solution. Further research and calculations have to give a physical explanation for the observed effect. The influence of heat transfer and diffusion on damping of free oscillations of a vapor bubble binary solution.

It was found that the dependence of the damping rate of oscillations of a bubble of water solutions of ethanol, methanol, and toluene monotonic on k_0.

It was mentioned for the aqueous solution of ethylene glycol similar dependence with a characteristic minimum at:

$$k_0 \approx 0,02.$$

Moreover, for:

$$0,01 \le k_0 \le 0,3$$

decrement, binary solution has less damping rates for pulsations of a bubble in pure (one-component) water and ethylene glycol.

This means that in the range of concentrations of water:

$$0,01 \le k_0 \le 0,3$$

Pulsations of the bubble (for water solution of ethylene glycol) decay much more slowly and there is inhibition of the process of phase transformations. A similar process was revealed and forced oscillations of bubbles in an acoustic field.

In this book the influence of nonstationary heat and mass transfer processes was investigated in the propagation of waves in a binary solution of liquids with bubbles. The influence of component composition and concentration of binary solution was investigated on the dispersion, dissipation and attenuation of monochromatic waves in two-phase, two-component media.

The aqueous solution of ethyl alcohol in aqueous ethylene glycol decrements showed perturbations less relevant characteristics of pure components of the solution.

Unsteady interphase heat transfer revealed in calculation, the structure of stationary shock waves in bubbly binary solutions. The problem signifies on effect a violation of monotonicity behavior of the calculated curves for concentration, indicating the presence of diffusion resistance.

In some of binary mixtures, it is seen the effect of diffusion resistance. It is led to inhibition of the intensity of phase transformations.

The physical explanation revealed the reason for an aqueous solution of ethylene glycol. Pronounced effect of diffusion resistance is related to the solution with limited ability. It diffuses through the components of $D = 10^{-9}$ (m^2/sec),

D – Diffusion coefficient volatility of the components is very different, and thus greatly different concentrations of the components in the solution and vapor phase.

In the case of aqueous solution of ethanol volatility component are roughly the same $\chi_1^0 \approx \chi_2^0$.

In accordance with Eq. (3) $c_0 \approx k_0$, so the finiteness of the diffusion coefficient does not lead to significant effects in violation of the thermal and mechanical equilibrium phases. Figures 4 and 5 show the dependence $k_0(c_0)$ of ethyl alcohol and ethylene glycol's aqueous solutions. From Fig.24 it is clear that almost the entire range of $k_0 \approx c_0$.

At the same time for an aqueous solution of ethylene glycol, by the calculations and Fig. 2 $0,01 \leq k_0 \leq 0,3$ $k_0 \leq c_0$, and when $k_0 > 0,3$ $k_0 \sim c_0$.

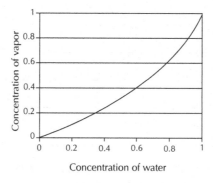

FIGURE 4   The dependence $(k_0 (c_0))$ for an aqueous solution of ethanol.

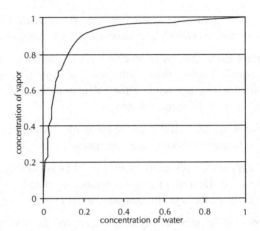

FIGURE 5   Dependence of $(k_0 (c_0))$ for an aqueous solution of ethylene glycol.

Figures 6 and 7 show the boiling point of the concentration for the solution of two systems.

When $k_0 = 1$, $c_0 = 1$ and get clean water to steam bubbles. It is for boiling of a liquid at $T_0 = 373°K$.

If $K_0 = 0$, $c_0 = 0$ and have correspondingly pure bubble ethanol ($T_0 = 373°K$) and ethylene glycol ($T_0 = 470°K$).

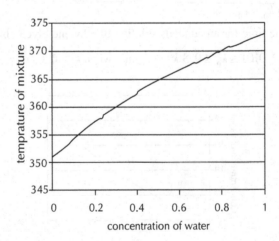

FIGURE 6   The dependence of the boiling temperature of the concentration of the solution to an aqueous solution of ethanol.

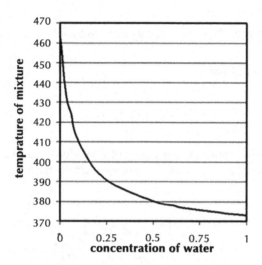

FIGURE 7 The dependence of the boiling point of the concentration of the solution to an aqueous solution of ethylene glycol.

It should be noted that in all the above works regardless of the above problems in the mathematical description of the cardinal effects of component composition of the solution shows the value of the parameter β equal:

$$\beta = \left(1 - \frac{1}{\gamma}\right) \frac{(c_0 - k_0)(N_{c_0} - N_{k_0})}{k_0(1 - k_0)} \frac{c_l}{c_{pv}} \left(\frac{c_{pv} T_0}{L}\right)^2 \sqrt{\frac{a_l}{D}}, \quad (19)$$

where,

N_{K_0}, N_{c_0} – molar concentration of 1-th component in the liquid and steam.

$$N_{k_0} = \frac{\mu k_0}{\mu k_0 + 1 - k_0},$$

$$N_{c_0} = \frac{\mu c_0}{\mu c_0 + 1 - c_0}$$

where,γ – Adiabatic index,

c_1 and c_{pv} respectively the specific heats of liquid and vapor at constant pressure,

a_l – Thermal diffusivity,

$$L = l_1 c_0 + l_2 \left(1 - c_0\right)$$

We also note that option (4) is a self-similar solution describing the growth of a bubble in a superheated solution. This solution has the form:

$$R = 2 \sqrt{\frac{3}{\pi}} \frac{\lambda_l \Delta T \sqrt{t}}{L \rho_v \sqrt{a_l \left(1 + \beta\right)}}, \tag{20}$$

Here ρ_v – vapor density,

t – time R – radius of the bubble,

λ_l is the coefficient of thermal conductivity,

ΔT – overheating of the liquid.

Figures 8 and 9 show the dependence $\beta(k_0)$ for the above binary solutions. For aqueous ethanol β is negative for any value of concentration and dependence on k_0 is monotonic.

For an aqueous solution of ethylene glycol β – is positive and has a pronounced maximum at $k_0 = 0,02$.

As a result of present work at low-pressure drops (superheating and super cooling of the liquid, respectively), diffusion does not occur in aqueous solutions of ethyl alcohol. By approximate equality of k_0 and c_0 all calculated dependence lie between the limiting curves for the case of one-component constituents of the solution.

They are included dependence of pressure, temperature, vapor bubble radius, the intensity of phase transformations, and so from time to time.

The pressure difference becomes important diffusion processes. Mass transfer between bubble and liquid is in a more intensive mode than in single-component constituents of the solution.

In particular, the growth rate of the bubble in a superheated solution is higher than in pure water and ethyl alcohol. It is because of the negative β according to Eq. (5).

In an aqueous solution of ethylene glycol, there is the same perturbations due to significant differences between k_0 and c_0. It is especially when $0,01 \le k_0 \le 0,3$, the effect of diffusion inhibition contributes to a significant intensity of mass transfer. In particular, during the growth of the bubble, the rate of growth in solution is much lower than in pure water and ethylene glycol.It is because of the positive β by Eq. (5).

Moreover, the maximum braking effect is achieved at the maximum value of β, when $k_0 = 0,02$.

A similar pattern is observed at the pulsations and the collapse of the bubble. Dependence of the damping rate of fluctuations in an aqueous solution of ethyl alcohol from the water concentration is monotonic. Aqueous solution of ethylene glycol dependence of the damping rate has a minimum at:

$$k_0 = 0,02.,0,01 \le k_0 \le 0,3.$$

The function decrement is small respectively large difference between k_0 and c_0 and β takes a large value. These ranges of concentrations in the solution have significant effect of diffusion inhibition.

For aqueous solutions of glycerin, methanol, toluene, etc., calculations are performed. Comparison with experimental data confirms the possibility of theoretical prediction of the braking of Heat and Mass Transfer.

It was analyzed the dependence of the parameter β, decrement of oscillations of a bubble from the equilibrium concentration of the mixture components. Therefore in every solutions, it was determined the concentration of the components of a binary mixture.

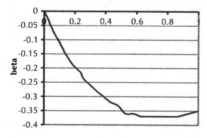

concentration of water

FIGURE 8   The dependence $\beta(k_0)$ for an aqueous solution of ethanol.

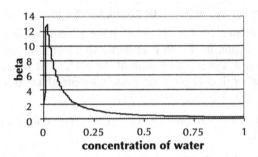

FIGURE 9 Dependence of $\beta(k_0)$ for an aqueous solution of ethylene glycol.

Figures 10 and 11 are illustrated by theoretical calculations. These figures defined on the example of aqueous solutions of ethyl alcohol and ethylene glycol (antifreeze used in car radiators). It is evident that the first solution is not suitable to the task.

The aqueous solution of ethylene glycol with a certain concentration is theoretically much more slowly boils over with clean water and ethylene glycol. This confirms the reliability of the method.

Calculations show that such a solution is almost never freezes. The same method can offer concrete solutions for cooling of hot parts and components of various machines and mechanisms.

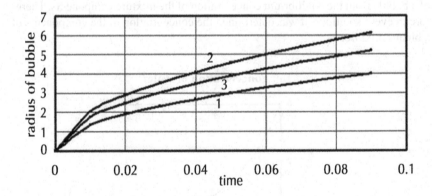

FIGURE 10 Dependence from time of vapor bubble radius. 1—water, 2—ethyl spirit, 3—water mixtures of ethyl spirit.

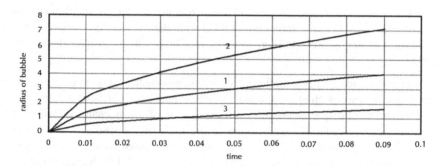

FIGURE 11   Dependence from time of vapor bubble radius.1—water, 2—ethylene glycol, 3—water mixtures of ethylene glycol.

The solution of the reduced system of equations revealed an interesting effect. The parameters were characterized the dynamics of bubbles in aqueous ethyl alcohol in the field of variable pressure. They lied between the limiting values of relevant parameters for the pure components. It was for the case, which pressure drops and consequently the role of diffusion was unimportant.

A completely different situation is observed during the growth and collapse of vapor bubble in aqueous solutions of ethylene glycol.The effect of diffusion resistance, leads to inhibition of the rate of phase transformations. For pure components of the solution, the growth and the collapse rate of the bubble is much smaller than the corresponding values.

4.3.8   THE STRUCTURE OF THE PRESSURE WAVE FOR A SIMPLE PIPELINE SYSTEM

The wave dynamics of dispersed two-phase mixtures, in contrast to homogeneous media, is determined by processes of interaction between phases. The essential difference between mechanical and physical-chemical properties of the phases leads to the fact that the external disturbance has on the carrier and the dispersed phase of different actions. As a result of the wave front of finite intensity phase is no longer in equilibrium and the resulting relaxation processes can significantly influence the course.

The rates of relaxation processes determine the structure of individual relaxation zones for elementary waves, and the whole flow as a whole.

For example, the rapid mechanical fragmentation of drops for shock and detonation waves leads to the formation of a large number of secondary droplets, the surface area that is several orders of magnitude higher than those for the source of aerosol.

That leads to rapid evaporation of the liquid, mixing the vapor on gaseous oxidizer and the formation of a homogeneous air-fuel mixture. The basis for studying the dynamics of disperses mixtures is a description of the mechanical motion of the mixture "in general" under review at the scale of the whole problem. Extensive interactions of the phases (including those caused by the deformation, fragmentation, evaporation and mixing), going at the scale of individual particles, determined by this macro scale motion, providing, in turn, a significant inverse effect on him.

Currently, conventional mathematical hydrodynamics model of heterogeneous environments is a model multispeed continuum, the most complete exposition in the works.

Investigate the structure of stationary shock waves in binary mixtures, propagating with the velocity of pressure wave. This speed is a fundamental parameter for modeling hydraulic transients. Consider the effect of the dynamics of vapor bubbles in gas-liquid two component mixtures on the propagation of shock waves, the effect of unsteady forces on transient flows, division of water columns in the vapor or steam bubbles, control steam pressure in cavities to pump up the direction of the pipeline. Solving this problem is extremely important to study all the conditions under which the piping system having adverse transients, in particular in the pumps and valves. In addition, it is important to develop methods of protection and devices to be used during the design and construction of separate parts of the system, as well as to identify their practical shortcomings.

In solving problems for the management of transients caused by abrupt change in pressure, suggests two possible strategies. The first strategy is to minimize the possibility of transient conditions during the design, determine the appropriate methods of flow control to eliminate the possibility of emergency and unusual situations in the system. The second strategy is to provide for the establishment of security device to control the possible transients due to events beyond control, such as equipment failures and power supply.

The Figs. 12 and 13 show the curves of pressure of time, received both experimental and theoretical methods. It is seen that the results of theoretical calculations agree well with experimental data.

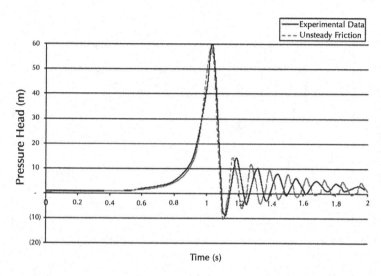

FIGURE 12   The structure of the pressure wave for a simple pipeline system using the model of equilibrium and no equilibrium friction.

In the experiments, the pipeline was equipped with a valve at the end of the main pipes, joined by timers to record time of closing. The characteristics of water hydraulic impact measured and recorded by extensometers in the computer memory. Pumping of water to the system was carried out by the pool, which allowed the pressure to stabilize the inlet.

The experiments were performed for three cases:

1.   A simple positive hydraulic shock for a straight pipe of constant diameter. The measured characteristics were the basis for assessing the impact of changes in diameter and the local diversion to the distribution of water hammer.

2.   Water hammer in pipelines with diameters change: contraction and expansion.

3.   Water hammer in pipelines with a local diversion in two scenarios: the outflow from the brain to super compression pool and a free outflow from the brain (to atmospheric pressure, with the ability to absorb air in the negative phase). This was the main reason for the air intake in the negative phase.

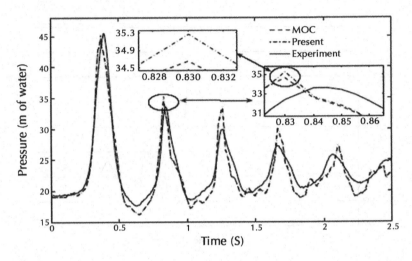

FIGURE 13   Comparison the results with experimental data.

Separation columns for pump power off on pipeline were carried out for two cases:

1. With the surge tank and the assumption of local diversion: In this case, the air was sucked into the pipeline,
2. Without the surge tank, provided the local diversion.

It can be recorded pressure wave and the wave velocity in fast transients up to 5 ms(in this study to 1 s). The assessment procedure was used to analyze Curve data, which were obtained on real systems. Software regression curve providing customized features and provided a regression analysis. Thus, the regression model (first model) was used in the final procedure. This model was compared with the results of the model performance (second model). The calculation results allowed us to develop a technical solution for the management of transition processes in the pipeline system. Figure 14 shows a scheme of how the developed device.

The surge tank with double bottom manages transitions, converting accumulated potential energy of water to kinetic energy. In critical situations during periods of rapid change in the nature of the flow of water from the reservoir flows into the system piping. The tank is usually located in the pumping station or highest point in the water system profile. In the vertical jumpers reservoir can be drilled various holes to control the supply of water from the system into the reservoir. But this design results in very little loss of flux, spilling from the tank. If there is overflow and leak, the tank can also act as a protective device against unintentional pressure increase.

Thus, the device may serve as a way of protection from the overflow. There is another important problem. The problem is related to stability in the reservoir. It is necessary for avoiding rapid rising or lowering the water level in the reservoir.

Therefore, the surface area of the tank must be much greater cross section of the pipeline [70-90].

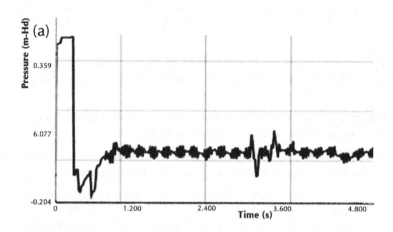

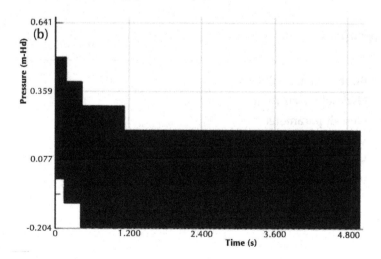

FIGURE 14   The working of the developed device for water supply.

4.4 CONCLUSION

It was evaluated by a comprehensive approach the strength of pipeline systems consisting of technical diagnostics segment of the pipe and the mathematical modeling of the pipeline. Results implicated on prediction of the selected model.

It was used a mathematical model for determining the nature of pulse propagation of pressure in pipelines. It was concluded the presence of bends, and local resistance. It gives a clear idea about the nature of pulse propagation of pressure in pipelines.

The experimental setup and performed on studies confirmed the validity of the proposed mathematical model, and comparing the results of calculations and experiments showed satisfactory agreement.

Industrial testing method for detection of water hammer for the water pipeline showed the effectiveness of the proposed methods.

It was confirmed the two temperature model the oretically on anomalous braking speed of phase transformations in boiling of binary mixtures.

It was investigated a comparison on accuracy of the numerical methods, regression model and the method of characteristics for the analysis of transient flows. It was shown that the method of characteristics is computationally more efficient for the analysis of large water pipelines.

KEYWORDS

- **Bubble binary solution**
- **Hydraulic transition**
- **Nusselt parameter**
- **Vapor bubble**
- **Vena contraction**
- **Water hammer**

REFERENCES

1. Qu, W.; Mala, G. M.; Li, D.; Heat Transfer for Water Flow in Trapezoidal Silicon Microchannels, **1993**, 399–404.
2. Hariri Asli, K.; Nagiyev, F. B.; Haghi, A. K.; Aliyev, S. A.; A computational approach to study fluid movement, 1st Festival on Water and Wastewater Research and Technology, Tehran, Iran, 12–17 Dec. **2009**, 27–32, http://isrc.nww.co.ir.
3. Peng, X. F.; Peterson, G. P.; Convective Heat Transfer and Flow Friction for Water Flow in Microchannel Structure, Int. J.; Heat Mass Transfer **1996**, *36,* 2599–2608.
4. Bergant Anton, Discrete Vapour Cavity Model with Improved Timing of Opening and Collapse of Cavities, **1980**, 1–11.
5. Ishii M.; Thermo-Fluid Dynamic Theory of Two-Phase Flow, Collection de D. R.; Liles and Reed, W. H.; "A Sern-Implict Method for Two-Phase Fluid la Direction des Etudes et. Recherché d'Electricite de France, 22 Dynamics," Journal of Computational Physics *26,* Paris, **1975**, 390–407.
6. Hariri Asli, K.; Nagiyev, F. B.; Haghi, A. K.; Aliyev, S. A.; A computational approach to study fluid movement, 1st Festival on Water and Wastewater Research and Technology, Tehran, Iran, 12–17 Dec. **2009**, 27–32, http://isrc.nww.co.ir.
7. Pickford J.; "Analysis of Surge," Macmillan, London, **1969**, 153–156.
8. Pipeline Design for Water and Wastewater, American Society of Civil Engineers, New York, **1975**, 54 p.
9. Xu B.; Ooi, K. T.; Mavriplis, C.; Zaghloul, M. E.; Viscous dissipation effects for liquid flow in microchannels, Micorsystems, **2002**, 53–57.
10. Fedorov, A. G.; Viskanta, R.; Three-dimensional Conjugate Heat Transfer into Microchannel Heat Sink for Electronic Packaging, *Int.J. Heat Mass Transfer* **2000**, *43,* 399–415.
11. Tuckerman, D. B.; Heat transfer microstructures for integrated circuits, Ph.D. thesis, Stanford University, **1984**, 10–120.
12. Harms, T. M.; Kazmierczak, M. J.; Cerner, F. M.; Holke A.; Henderson, H. T.; Pilchowski, H. T.; Baker, K.; Experimental Investigation of Heat Transfer and Pressure Drop through Deep Micro channels in a (100) Silicon Substrate, in: Proceedings of the ASME.; Heat Transfer Division, HTD **1997**, *351,* 347–357.
13. Holland, F. A.; Bragg, R.; Fluid Flow for Chemical Engineers, Edward Arnold Publishers, London, **1995**, 1–3.
14. Lee, T. S.; Pejovic, S. Air influence on similarity of hydraulic transients and vibrations. *ASME J. Fluid Eng.* **1996**, *118(4),* 706–709.
15. Li J.; McCorquodale A.; "Modeling Mixed Flow in Storm Sewers," Journal of Hydraulic Engineering, ASCE, **1999**, *125(11),* 1170–1180.
16. Minnaert M.; on musical air bubbles and the sounds of running water. Phil. Mag.; **1933**, *16(7),* 235–248.
17. Moeng, C. H.; McWilliams, J. C.; Rotunno, R.; Sullivan, P. P.; Weil J.; "Investigating 2D modeling of atmospheric convection in the PBL," *J. Atm. Sci.* **2004**, *61,* 889 −903.
18. Tuckerman, D. B.; Pease, R. F. W. High performance heat sinking for VLSI, IEEE Electron device letter, DEL-2, **1981**, 126–129.

19. Nagiyev, F. B.; Khabeev, N. S, Bubble dynamics of binary solutions. High Temperature, **1988**, *27(3),* 528–533.

20. Shvarts D.; Oron D.; Kartoon D.; Rikanati A.; Sadot O.; "Scaling laws of nonlinear Rayleigh-Taylor and Richtmyer-Meshkov instabilities in two and three dimensions,"C.R.Acad. Sci. Paris, IV; *719,* **2000,** 312 p.

21. Cabot, W. H.; Cook, A. W.; Miller, P. L.; Laney, D. E.; Miller, M. C.; Childs, H. R.; "Large eddy simulation of Rayleigh-Taylor instability," Phys. Fluids, September, **2005,** *17,* 91–106.

22. Cabot W.; University of California, Lawrence Livermore National laboratory, Livermore, CA, *Phys. Fluids,* **2006,** 94–550.

23. Goncharov V. N.; "Analytical model of nonlinear, single-mode, classical Rayleigh-Taylor instability at arbitrary Atwood numbers,"*Phys. Rev. Lett.* **2002,** *88,* 134502, 10–15.

24. Ramaprabhu P.; Andrews, M. J.; "Experimental investigation of Rayleigh-Taylor mixing at small Atwood numbers,"*J. Fluid Mech. 502,* **2004,** 233 p.

25. Clark, T. T.; "A numerical study of the statistics of a two-dimensional Rayleigh-Taylor mixing layer,"*Phys. Fluids* **2003,** *15,* 2413.

26. Cook, A. W.; Cabot W.; Miller, P. L.; "The mixing transition in Rayleigh-Taylor instability,"*J. Fluid Mech. 511,* **2004,** 333.

27. Waddell, J. T.; Niederhaus, C. E.; Jacobs, J. W.; "Experimental study of Rayleigh-Taylor instability: Low Atwood number liquid systems with single-mode initial perturbations,"*Phys. Fluids 13,* **2001,** 1263–1273.

28. Weber, S. V.; Dimonte G.; Marinak, M. M.; "Arbitrary Lagrange-Eulerian code simulations of turbulent Rayleigh-Taylor instability in two and three dimensions," Laser and Particle Beams **2003,** *21,* 455 p.

29. Dimonte G.; Youngs D.; Dimits A.; Weber S.; Marinak M. "A comparative study of the Rayleigh-Taylor instability using high-resolution three-dimensional numerical simulations: the Alpha group collaboration,"*Phys. Fluids* **2004,** *16,* 1668.

30. Young, Y. N.; Tufo, H.; Dubey, A.; Rosner, R.; "On the miscible Rayleigh-Taylor instability: two and three dimensions,"*J. Fluid Mech.* **2001,** *447,* 377, 2003–2500.

31. George, E.; Glimm, J.; "Self-similarity of Rayleigh-Taylor mixing rates,"*Phys. Fluids 17,* 054101, **2005,** 1–3.

32. Oron, D.; Arazi, L.; Kartoon D.; Rikanati A.; Alon U.; Shvarts D.; "Dimensionality dependence of the Rayleigh-Taylor and Richtmyer-Meshkov instability late-time scaling laws,"*Phys. Plasmas* **2001,** *8,* **2883,**

33. Nigmatulin, R. I.; Nagiyev, F. B.; Khabeev, N. S.; Effective heat transfer coefficients of the bubbles in the liquid radial pulse. Mater. Second-Union. Conf. Heat Mass Transfer, "Heat massoob-men in the biphasic with Minsk,"**1980,** *5,* 111–115.

34. Nagiyev, F. B.; Khabeev N. S, Bubble dynamics of binary solutions. High Temperature, **1988,** *27(3),* 528–533.

35. Nagiyev, F. B.; Damping of the oscillations of bubbles boiling binary solutions. Mater. VIII Resp. Conf. mathematics and mechanics. Baku, October 26–29, **1988,** 177–178.

36. Nagyiev, F. B.; Kadyrov, B. A.; Small oscillations of the bubbles in a binary mixture in the acoustic field. Math. An Az. SSR Ser. Physicotech. Mate. *Science,* **1986,** *1,* 23–26.

37. Nagiyev, F. B.; Dynamics, heat and mass transfer of vapor-gas bubbles in a two-component liquid. Turkey-Azerbaijan petrol semin.; Ankara, Turkey, **1993**, 32–40.

38. Nagiyev, F. B.; The method of creation effective coolness liquids, Third Baku international Congress. Baku, Azerbaijan Republic, **1995**, 19–22.

39. Nagiyev, F. B.; The linear theory of disturbances in binary liquids bubble solution. Dep. In VINITI, **1986**, *405(86)*, 76–79.

40. Nagiyev, F. B.; Structure of stationary shock waves in boiling binary solutions. Math. USSR, Fluid Dynamics, **1989**, *1*, 81–87.

41. Rayleigh, On the pressure developed in a liquid during the collapse of a spherical cavity. Philos. Mag. Ser.6, **1917**, *34(200)*, 94–98.

42. Perry, R. H.; Green, D. W.; Maloney, J. O.; Perry's Chemical Engineers Handbook, 7th Edition, McGraw-Hill, New York, **1997**, 1–61.

43. Nigmatulin, R. I.; Dynamics of multiphase media. Moscow, Nauka, **1987**, *1(2)*, 12–14.

44. Kodura A.; Weinerowska K.; the influence of the local pipeline leak on water hammer properties, Materials of the II Polish Congress of Environmental Engineering, Lublin, **2005**, 125–133.

45. Kane J.; Arnett D.; Remington, B. A.; Glendinning, S. G.; Baz'an G.; "Two-dimensional versus three-dimensional supernova hydrodynamic instability growth," Astrophys. J.; **2000**, 528–989.

46. Quick, R. S.; "Comparison and Limitations of Various Water hammer Theories,"*J. Hyd. Div. ASME*, May, **1933**, 43–45.

47. Jaeger C.; "Fluid Transients in Hydro-Electric Engineering Practice," Blackie and Son Ltd.; **1977**, 87–88.

48. Jaime Suárez A.; "Generalized water hammer algorithm for piping systems with unsteady friction" **2005**, 72–77.

49. Fok, A.; Ashamalla A.; Aldworth G.; "Considerations in Optimizing Air Chamber for Pumping Plants," Symposium on Fluid Transients and Acoustics in the Power Industry, San Francisco, USA, Dec, **1978**, 112–114.

50. Fok, A.; "Design Charts for Surge Tanks on Pump Discharge Lines," BHRA 3rd Int. Conference on Pressure Surges, Bedford, England, Mar.; **1980**, 23–34.

51. Fok, A.; "Water hammer and Its Protection in Pumping Systems," Hydro technical Conference, CSCE, Edmonton, May, **1982**, 45–55.

52. Fok, A.; "A contribution to the Analysis of Energy Losses in Transient Pipe Flow," PhD.; Thesis, University of Ottawa, **1987**, 176–182.

53. Hariri Asli, K.; Nagiyev, F. B.; Water Hammer and fluid condition, Ministry of Energy, Gilan Water and Wastewater Co.; Research Week Exhibition, Tehran, Iran, December, **2007**, 132–148, http://isrc.nww.co.ir.

54. Hariri Asli, K.; Nagiyev, F. B.; Water Hammer analysis and formulation, Ministry of Energy, Gilan Water and Wastewater Co.; Research Week Exhibition, Tehran, Iran, December, **2007**, 111–131, http://isrc.nww.co.ir.

55. Hariri Asli, K.; Nagiyev, F. B.; Water Hammer and hydrodynamics instabilities, Interpenetration of two fluids at parallel between plates and turbulent moving in pipe, Ministry of Energy, Guilan Water and Wastewater Co.; Research Week Exhibition, Tehran, Iran, December, **2007**, 90–110, http://isrc.nww.co.ir.

56. Hariri Asli, K.; Nagiyev, F. B.; Water Hammer and pump pulsation, Ministry of Energy, Guilan Water and Wastewater Co.; Research Week Exhibition, Tehran, Iran, December, **2007**, 51–72, http://isrc.nww.co.ir.

57. Hariri Asli, K.; Nagiyev, F. B.; Reynolds number and hydrodynamics' instability," Ministry of Energy, Guilan Water and Wastewater Co.; Research Week Exhibition, Tehran, Iran, December, **2007**, 31–50, http://isrc.nww.co.ir.

58. Hariri Asli, K.; Nagiyev, F. B.; Water Hammer and valves, Ministry of Energy, Guilan Water and Wastewater Co.; Research Week Exhibition, Tehran, Iran, December, **2007**, 20–30, http://isrc.nww.co.ir.

59. Hariri Asli, K.; Nagiyev, F. B.; "Interpenetration of two fluids at parallel between plates and turbulent moving in pipe," Ministry of Energy, Guilan Water and Wastewater Co.; Research Week Exhibition, Tehran, Iran, December, **2007**, 73–89, http://isrc.nww.co.ir.

60. Hariri Asli, K.; Nagiyev, F. B.; Decreasing of Unaccounted For Water "UFW" by Geographic Information System"GIS" in Rasht urban water system, civil engineering organization of Guilan, Technical and Art Journal, **2007**, 3–7, http://www.art-of-music.net/.

61. Hariri Asli, K.; Portable Flow meter Tester Machine Apparatus, Certificate on registration of invention, Tehran, Iran, #010757, Series a/82, 24/11/2007, 1–3.

62. Hariri Asli, K.; Nagiyev, F. B.; Haghi, A. K.; "Interpenetration of two fluids at parallel between plates and turbulent moving in pipe," 9th Conference on Ministry of Energetic works at research week, Tehran, Iran, **2008**, 73–89, http://isrc.nww.co.ir.

63. Hariri Asli, K.; Nagiyev, F. B.; Haghi, A. K.; "Water hammer and valves," 9th Conference on Ministry of Energetic works at research week, Tehran, Iran, **2008**, 20–30, http://isrc.nww.co.ir.

64. Hariri Asli, K.; Nagiyev, F. B.; Haghi, A. K.; "Water hammer and hydrodynamics instability," 9th Conference on Ministry of Energetic works at research week, Tehran, Iran, **2008**, 90–110, http://isrc.nww.co.ir.

65. Hariri Asli, K.; Nagiyev, F. B.; Haghi, A. K.; "Water hammer analysis and formulation," 9th Conference on Ministry of Energetic works at research week, Tehran, Iran, **2008**, 27–42, http://isrc.nww.co.ir.

66. Hariri Asli, K.; Nagiyev, F. B.; Haghi, A. K.; "Water hammer & fluid condition," 9th Conference on Ministry of Energetic works at research week, Tehran, Iran, **2008**, 27–43, http://isrc.nww.co.ir.

67. Hariri Asli, K.; Nagiyev, F. B.; Haghi, A. K.; "Water hammer and pump pulsation," 9th Conference on Ministry of Energetic works at research week, Tehran, Iran, **2008**, 27–44, http://isrc.nww.co.ir.

68. Hariri Asli, K.; Nagiyev, F. B.; Haghi, A. K.; "Reynolds number and hydrodynamics instability," 9th Conference on Ministry of Energetic works at research week, Tehran, Iran, **2008**, 27–45, http://isrc.nww.co.ir.

69. Hariri Asli, K.; Nagiyev, F. B.; Haghi, A. K.; "Water hammer and fluid Interpenetration," 9th Conference on Ministry of Energetic works at research week, Tehran, Iran, **2008**, 27–47, http://isrc.nww.co.ir.

70. Hariri Asli, K.; GIS and water hammer disaster at earthquake in Rasht water pipeline, civil engineering organization of Guilan, Technical and Art Journal, **2008**, 14–17, http://www.art-of-music.net/.

71. Hariri Asli, K.; GIS and water hammer disaster at earthquake in Rasht water pipeline, 3rd International Conference on Integrated Natural Disaster Management, Tehran university, ISSN: 1735–5540, 18–19 Feb.; INDM, Tehran, Iran, **2008**, *13,* 53/1–12, http://www.civilica.com/Paper-INDM03-INDM03_001.html

72. Hariri Asli, K.; Nagiyev, F. B.; Bubbles characteristics and convective effects in the binary mixtures. Transactions issue mathematics and mechanics series of physical-technical and mathematics science, ISSN: 0002–3108, Azerbaijan, Baku, **2009**, 68–74, http://www.imm.science.az/journals.html.

73. Hariri Asli, K.; Nagiyev, F. B.; Haghi, A. K.; Aliyev, S. A.; Three-Dimensional conjugate heat transfer in porous media, 1st Festival on Water and Wastewater Research and Technology, Tehran, Iran, 12–17 Dec. **2009**, 26–28, http://isrc.nww.co.ir.

74. Hariri Asli, K.; Nagiyev, F. B.; Haghi, A. K.; Aliyev, S. A.; Some Aspects of Physical and Numerical Modeling of water hammer in pipelines, 1st Festival on Water and Wastewater Research and Technology, Tehran, Iran, 12–17 Dec. **2009**, 26–29, http://isrc.nww.co.ir

75. Hariri Asli, K.; Nagiyev, F. B.; Haghi, A. K.; Aliyev, S. A.; Modeling for Water Hammer due to valves: From theory to practice, 1st Festival on Water and Wastewater Research and Technology, Tehran, Iran, 12–17 Dec. **2009**, 26, 30, http://isrc.nww.co.ir.

76. Hariri Asli, K.; Nagiyev, F. B.; Haghi, A. K.; Aliyev, S. A.; Water hammer and hydrodynamics instabilities modeling: From Theory to Practice, 1st Festival on Water and Wastewater Research and Technology, Tehran, Iran, 12–17 Dec. **2009**, 26–31, http://isrc.nww.co.ir

77. Hariri Asli, K.; Nagiyev, F. B.; Haghi, A. K.; Aliyev, S. A.; A computational approach to study fluid movement, 1st Festival on Water and Wastewater Research and Technology, Tehran, Iran, 12–17 Dec.2009, 27–32, http://isrc.nww.co.ir.

78. Hariri Asli, K.; Nagiyev, F. B.; Haghi, A. K.; Aliyev, S. A.; Water Hammer Analysis: Some Computational Aspects and practical hints, 1st Festival on Water and Wastewater Research and Technology, Tehran, Iran, 12–17 Dec. **2009**, 27–33, http://isrc.nww.co.ir

79. Hariri Asli, K.; Nagiyev, F. B.; Haghi, A. K.; Aliyev, S. A.; Water Hammer and Fluid condition: A computational approach, 1st Festival on Water and Wastewater Research and Technology, Tehran, Iran, 12–17 Dec. **2009**, 27–34, http://isrc.nww.co.ir.

80. Hariri Asli, K.; Nagiyev, F. B.; Haghi, A. K.; Aliyev, S. A.; A computational Method to Study Transient Flow in Binary Mixtures, 1st Festival on Water and Wastewater Research and Technology, Tehran, Iran, 12–17 Dec. **2009**, 27–35, http://isrc.nww.co.ir.

81. Hariri Asli, K.; Nagiyev, F. B.; Haghi, A. K.; Physical modeling of fluid movement in pipelines, 1st Festival on Water and Wastewater Research and Technology, Tehran, Iran, 12–17 Dec. **2009**, 27–36, http://isrc.nww.co.ir.

82. Hariri Asli, K.; Nagiyev, F. B.; Haghi, A. K.; Aliyev, S. A.; Interpenetration of two fluids at parallel between plates and turbulent moving, 1st Festival on Water and Wastewater Research and Technology, Tehran, Iran, 12–17 Dec. **2009**, 27–37, http://isrc.nww.co.ir.

83. Hariri Asli, K.; Nagiyev, F. B.; Haghi, A. K.; Aliyev, S. A.; Modeling of fluid interaction produced by water hammer, 1st Festival on Water and Wastewater Research and Technology, Tehran, Iran, 12–17 Dec. **2009,** 27–38, http://isrc.nww.co.ir.

84. Hariri Asli, K.; Nagiyev, F. B.; Haghi, A. K.; Aliyev, S. A.; GIS and water hammer disaster at earthquake in Rasht pipeline, 1st Festival on Water and Wastewater Research and Technology, Tehran, Iran, 12–17 Dec. **2009,** 27–39, http://isrc.nww.co.ir.

85. Hariri Asli, K.; Nagiyev, F. B.; Haghi, A. K.; Aliyev, S. A.; Interpenetration of two fluids at parallel between plates and turbulent moving, 1st Festival on Water and Wastewater Research and Technology, Tehran, Iran, 12–17 Dec. **2009,** 27–40, http:// isrc.nww.co.ir.

86. Hariri Asli, K.; Nagiyev, F. B.; Haghi, A. K.; Aliyev, S. A.; Water hammer and hydrodynamics' instability, 1st Festival on Water and Wastewater Research and Technology, Tehran, Iran, 12–17 Dec. **2009,** 27–41, http://isrc.nww.co.ir.

87. Hariri Asli, K.; Nagiyev, F. B.; Haghi, A. K.; Aliyev, S. A.; Water hammer analysis and formulation, 1st Festival on Water and Wastewater Research and Technology, Tehran, Iran, 12–17 Dec. **2009,** 27–42, http://isrc.nww.co.ir.

88. Hariri Asli, K.; Nagiyev, F. B.; Haghi, A. K.; Aliyev, S. A.; Water hammer & fluid condition, 1st Festival on Water and Wastewater Research and Technology, Tehran, Iran, 12–17 Dec. **2009,** 27–43, http://isrc.nww.co.ir.

89. Hariri Asli, K.; Nagiyev, F. B.; Haghi, A. K.; Aliyev, S. A.; Water hammer and pump pulsation, 1st Festival on Water and Wastewater Research and Technology, Tehran, Iran, 12–17 Dec. **2009,** 27–44, http://isrc.nww.co.ir.

90. Hariri Asli, K.; Nagiyev, F. B.; Haghi, A. K.; Aliyev, S. A.; Reynolds number and hydrodynamics instabilities, 1st Festival on Water and Wastewater Research and Technology, Tehran, Iran, 12–17 Dec. 2009, 27–45, http://isrc.nww.co.ir.

CHAPTER 5

MATHEMATICAL CONCEPTS AND COMPUTATIONAL APPROACH ON HYDRODYNAMICS INSTABILITY

CONTENTS

5.1   INTRODUCTION

In this book a case study with computational approach on hydrodynamics instability for water pipeline was presented. A nonlinear heterogeneous model was defined for investigation the flow which interred into the surge tank. The proportionality of this flow to total discharge in the pipeline was affected by the values of oscillations period. In a consequence it was affected to the value of wave celerity.

N. E. Zhukovsky introduced the concept of the effective sound speed. He mentioned to reducing the motion of a compressible fluid in an elastic cylindrical pipe to the motion of a compressible fluid in a rigid pipe, but with a lower modulus of elasticity of the liquid. Calculations of hydraulic shock in multiphase systems, including a computer, are devoted to the work of V. M. Alysheva. In that work, integration of differential equations of unsteady pressure flow is also performed by the "method of characteristics." The works of Streater, K. P. Vishnevsky, B.F. Lyamaeva, and V. M. Alyshev use the method of calculation of water hammer. They are based on replacing the distributed along the length of the flow of gas parameters concentrated in the fictitious air-hydraulic caps installed on the boundaries of the pipeline. A fictitious elastic element is replaced by elastic deformation of the pipe walls, and the elastic deformation of the solid suspension is modeled by fictitious elastic elements of the solid suspension. However, detailed experimental studies are based on the solid component [1-19].

5.2   MATERIALS AND METHODS

The influence of the ratio of total discharge in the pipeline was affected by the values of oscillations period. In a consequence it was affected to the value of wave celerity if the outflow to the overpressure reservoir from the leak was imposed.

A particular challenge in terms of calculations is a hydraulic shock, accompanied by discontinuities of the flow. This phenomenon has not been fully studied, so the works of many scientists are devoted to experimental and theoretical studies of nonstationary processes with a break of the column of liquid.

First detailed study and writing the first design formula, for such cases of hydraulic shocks with discontinuities of continuous flow, was the work of A. F. Masty. Subsequently, the development of this issue was paid more attention by A. Surin, L. Bergeron, L. F. Moshnin, N. A. Kartvelishvili, M. Andriashev, V. S.

Dikarevsky, K. P. Wisniewski, B. F. Laman, V. I. Blokhin, L. S. Gerashchenko, V. N. Kovalenko, and others.

The most detailed experimental and theoretical study of water hammer with a discontinuity in the flow conduits performed by D. N. Smirnov and L. B. Zubov. As a result of the research, they describe the basic laws of gap columns, fluid and obtained relatively simple calculation dependences. In the above works, there are methods of determining maximal pressures after the discontinuities of the flow. However, the results of calculations by these methods are often contradictory. In addition, not clarified the conditions under which the maximum pressure generated. There is little influence of loss of pressure, vacuum, nature and duration of flow control and other factors on the value of maximum pressure.

The study of V. S. Dikarevsky for water hammer was included to break the continuity of flow. His work dealt with in detail, the impact magnitude of the vacuum on the course of the entire process of water hammer. Analytically and based on experimental data, scholars have argued that in a horizontal pipe rupture. The continuity of the flow occurs mainly in the regulatory body, and cavitation phenomena on the length of the pipeline are manifested. It investigates only in the form of small bubbles, whose influence on the process of hydraulic impact is negligible. As a result, research scientists have obtained analytic expressions for the hydraulic shock. They mention a gap of continuous flow, taking into account the energy loss, while controlling the flow and the wave nature. However, studies of V. S. Dikarevskogo were conducted mainly for the horizontal pressure pipelines and pumping units with a low inertia of moving masses.

Researches of N. I. Kolotilo and others devoted to the study of water hammer to break the continuity of flow in the intermediate point. N. I. Kolotilo analytically derived the condition for the gap of continuous flow at a turning point of the pipeline when the pressure is reduced at this point (below atmospheric pressure). Development of algorithms for software simulation of transients by K. P. Vishnevsky was made for the complex pressure systems. It included the possible formation of discontinuities flows, hydraulic resistance, structural features of the pumping of water systems [20-44].

5.3   RESULTS

Calculation of water hammer is adapted to high-pressure water systems for household and drinking purposes. K. P. Vishnevsky used "characteristics method" for the calculation of water hammer on a computer dedicated to the work of B. F. Lyamaeva. They described in detail the process of modeling the unsteady fluid flow in complex piping systems transporting drinking water.

Their works were included the description of this phenomenon at discontinuities flow, unsteady friction, changes in gas content and other parameters [45-78].

This work led to improved standards for precession designs and installation techniques in the field of Sub atmospheric transient pressures into the water system.

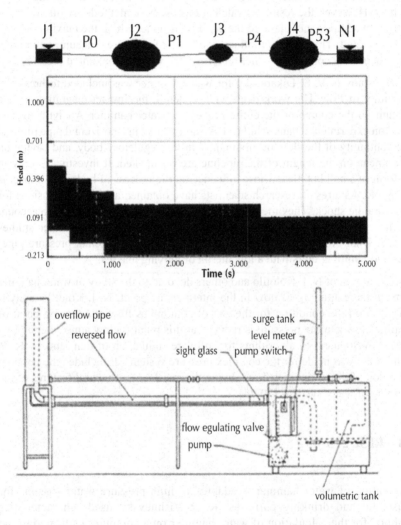

FIGURE 1   Laboratory model results (present work).

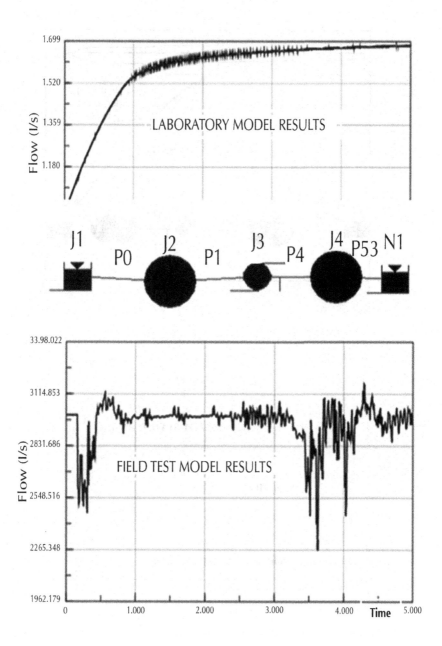

FIGURE 2 Laboratory experiments for flow and pressure data records.

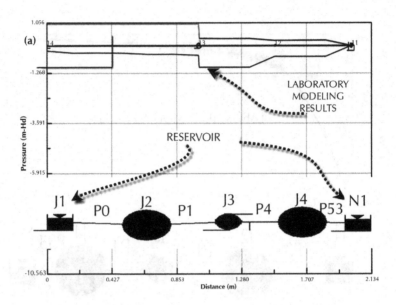

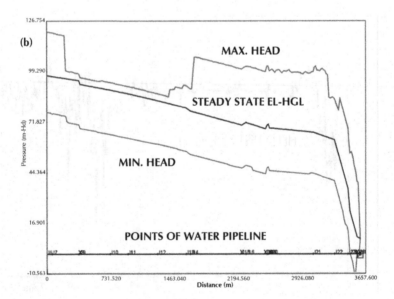

FIGURE 3 (a) Laboratory experiments for data records, (b) Water Pipeline equipped with surge tank.

5.4   CONCLUSION

Studies have shown that the location of the discontinuity of continuous flow at a turning point depends, first of all, the profile of the pipeline. Protection of hydraulic systems against water hammer by releasing part of the transported fluid is the most widespread method of artificial reduction of the hydraulic shock. Devices that perform this function can be divided into valve, bursting disc and the overflow of the column.

KEYWORDS

- **Computational method**
- **Hydraulic transient modeling**
- **Hydraulic transition**
- **Structural features**
- **Water hammer**
- **Water systems**

REFERENCES

1. Arturo Leon, S. Improved modeling of unsteady free surface, pressurized and mixed flows in storm-sewer systems, Submitted in partial fulfillment of the requirements for the degree of Doctor of Philosophy in Civil Engineering in the Graduate College of the University of Illinois at Urbana-Champaign, **2007,** 57–58,
2. Hariri Asli, K.; Nagiyev, F. B.; Haghi A. K. Computational methods in applied science and engineering. In: Interpenetration of Two Fluids at Parallel Between Plates and Turbulent Moving in Pipe. Nova Science, New York, USA, **2009,** 115–128 Chapter7, https://www.novapublishers.com/catalog/ product_ info.php? products_id=10681.
3. Wylie, E. B.; Streeter, V. L. Fluid transients, Feb Press, Ann. Arbor, MI, **1983,** corrected copy: **1982,** 166–171.
4. Apoloniusz, Kodura, Katarzyna, Weinerowska: Some aspects of physical and numerical modeling of Water Hammer in Pipelines, **2005,** 125–133.
5. Hariri Asli, K.; Nagiyev, F. B.; Haghi A. K. "Some Aspects of Physical and Numerical Modeling of water hammer in pipelines," Nonlinear DynamicsAn International Journal of Nonlinear Dynamics and Chaos in Engineering Systems, ISSN: 1573–269X

(electronic version) Journal no. 11071 Springer, Netherland, ISSN: 0924–090X (Print version), Germany, **2010**, *60(4)*, 677–701, June, http://www.springerlink.com/open-url.asp?genre=article&id=doi: 10.1007/s11071-009-9624-7.

6. Hariri Asli, K.; Nagiyev, F. B.; Haghi, A. K.; Aliyev, S. A. Physical and Numerical Modeling of Fluid Flow in Pipelines: A computational approach, International Journal of the Balkan Tribological Association, ISSN: 1310–4772, Sofia, Bulgaria, **2010**, *16(1)*, 20–34.

7. Hariri Asli, K.; Nagiyev, F. B.; Haghi, A. K.; Aliyev S. A. Nonlinear Heterogeneous Model for Water Hammer Disaster, International Journal of the Balkan Tribological Association, ISSN: 1310–4772, Sofia, Bulgaria, **2010**, *16(2)*, 209–222.

8. Hariri Asli, K.; Nagiyev, F. B.; Haghi, A. K.; Aliyev S. A, Hariri Asli H. Numerical modeling of transients flow in water pipeline: A computational approach, International Journal of Academic Research, ISSN: 1310–4772, Baku, Azerbaijan, ISSN: 2075–4124, **2010**, *2(5)*, September 30.

9. Lee, T. S.; Pejovic, S. Air influence on similarity of hydraulic transients and vibrations, *ASME J. Fluid Eng.* **1996**, *118(4)*, 706–709.

10. Fedorov, A. G.; Viskanta R.; Three-dimensional Conjugate Heat Transfer into Microchannel Heat Sink for Electronic Packaging, Int. J.; Heat Mass Transfer *43*, **2000**, 399–415.

11. Tuckerman, D. B.; Heat transfer microstructures for integrated circuits, Ph.D. thesis, Stanford University, **1984**, 10–120.

12. Harms, T. M.; Kazmierczak, M. J.; Cerner, F. M.; Holke, A.; Henderson, H. T.; Pilchowski, H. T.; Baker K.; Experimental Investigation of Heat Transfer and Pressure Drop through Deep Micro channels in a (100) Silicon Substrate, In: Proceedings of the ASME Heat Transfer Division, HTD**1997**, *351*, 347–357.

13. Holland, F. A.; Bragg R.; Fluid Flow for Chemical Engineers, Edward Arnold Publishers, London, **1995**, 1–3.

14. Lee, T. S.;Pejovic S Air influence on similarity of hydraulic transients and vibrations. *ASME J. Fluid Eng.* **1996**, *118(4)*,706–709.

15. Li J.; McCorquodale A.; "Modeling Mixed Flow in Storm Sewers," Journal of Hydraulic Engineering, ASCE, **1999**, *125(11)*, 1170–1180.

16. Minnaert M.;on musical air bubbles and the sounds of running water. Phil. Mag.; **1933**, *16(7)*, 235–248.

17. Moeng, C. H.; McWilliams, J. C.; Rotunno, R.; Sullivan, P. P.; Weil, J.; "Investigating 2D modeling of atmospheric convection in the PBL,"*J. Atm. Sci.* **2004**, *61*, 889 −903.

18. Tuckerman, D. B.; R. F. W Pease, high performance heat sinking for VLSI, IEEE Electron device letter, DEL-2, **1981**, 126–129.

19. Nagiyev, F. B.; Khabeev,N. S, Bubble dynamics of binary solutions. High Temperature, **1988**, v. *27(3)*, 528–533.

20. Shvarts D.; Oron D.; Kartoon D.; Rikanati A.; Sadot O.; "Scaling laws of nonlinear Rayleigh-Taylor and Richtmyer-Meshkov instabilities in two and three dimensions,"*C.R. Acad. Sci. Paris IV,* **2000**, *719*, 312 p.

21. Cabot, W. H.; Cook, A. W.; Miller, P. L.; Laney, D. E.; Miller, M. C.; Childs, H. R.; "Large eddy simulation of Rayleigh-Taylor instability," Phys. Fluids, September, **2005**, 17, 91–106.

22. Cabot W.; University of California, Lawrence Livermore National laboratory, Livermore, CA, *Phys. Fluids,* **2006**, 94–550.
23. Goncharov V. N.; "Analytical model of nonlinear, single-mode, classical Rayleigh-Taylor instability at arbitrary Atwood numbers,"*Phys. Rev. Lett.* **2002**, *88,* 134502, 10–15.
24. Ramaprabhu P.; Andrews, M. J.; "Experimental investigation of Rayleigh-Taylor mixing at small Atwood numbers,"*J. Fluid Mech.* **2004**, *502,* 233 p.
25. Clark, T. T.; "A numerical study of the statistics of a two-dimensional Rayleigh-Taylor mixing layer,"*Phys. Fluids* **2003**, *15,* 2413.
26. Cook, A. W.; Cabot W.; Miller, P. L.; "The mixing transition in Rayleigh-Taylor instability,"*J. Fluid Mech.* **2004**, *511,* 333.
27. Waddell, J. T.; Niederhaus, C. E.; Jacobs, J. W.; "Experimental study of Rayleigh-Taylor instability: Low Atwood number liquid systems with single-mode initial perturbations,"*Phys. Fluids* **2001**, *13,* 1263–1273.
28. Weber, S. V.; Dimonte, G.; Marinak, M. M.; "Arbitrary Lagrange-Eulerian code simulations of turbulent Rayleigh-Taylor instability in two and three dimensions," Laser and Particle Beams **2003**, *21,* 455 p.
29. Dimonte G.; Youngs D.; Dimits A.; Weber S.; Marinak M. "A comparative study of the Rayleigh-Taylor instability using high-resolution three-dimensional numerical simulations: the Alpha group collaboration,"*Phys. Fluids* **2004**, *16,* 1668.
30. Young, Y. N.; Tufo, H.; Dubey, A.; Rosner, R.; "On the miscible Rayleigh-Taylor instability: two and three dimensions,"*J. Fluid Mech.* **2001**, *447, 377,* 2003–2500.
31. George E.; Glimm J.; "Self-similarity of Rayleigh-Taylor mixing rates,"*Phys. Fluids* **2005**, *17,* 054101, 1–3.
32. Oron, D.; Arazi, L.; Kartoon, D.; Rikanati, A.; Alon, U.; Shvarts, D.; "Dimensionality dependence of the Rayleigh-Taylor and Richtmyer-Meshkov instability late-time scaling laws,"*Phys. Plasmas* **2001**, *8,* 2883.
33. Nigmatulin, R. I.; Nagiyev, F. B.; Khabeev, N. S.; Effective heat transfer coefficients of the bubbles in the liquid radial pulse. Mater. Second-Union. Conf. Heat Mass Transfer, "Heat massoob-men in the biphasic with Minsk,"**1980**, *5,* 111–115.
34. Nagiyev, F. B.; Khabeev, N. S, Bubble dynamics of binary solutions. High Temperature, **1988**, *27(3),* 528–533.
35. Nagiyev, F. B.; Damping of the oscillations of bubbles boiling binary solutions. Mater. VIII Resp. Conf. mathematics and mechanics. Baku, October 26–29, **1988**, 177–178.
36. Nagyiev, F. B.; Kadyrov, B. A.; Small oscillations of the bubbles in a binary mixture in the acoustic field. Math. AN Az.SSR Ser. Physicotech. Mate. Science, **1986**, *1,* 23–26.
37. Nagiyev, F. B.;Dynamics, heat and mass transfer of vapor-gas bubbles in a two-component liquid. Turkey-Azerbaijan petrol semin.;Ankara, Turkey, **1993**, 32–40.
38. Nagiyev, F. B.; The method of creation effective coolness liquids, Third Baku international Congress. Baku, Azerbaijan Republic, **1995**, 19–22.
39. Nagiyev, F. B.; The linear theory of disturbances in binary liquids bubble solution. Dep. In VINITI, **1986**, *405, 86,* 76–79.
40. Nagiyev, F. B.; Structure of stationary shock waves in boiling binary solutions. Math. USSR, Fluid Dynamics, **1989**, *1,* 81–87.

41. Rayleigh, On the pressure developed in a liquid during the collapse of a spherical cavity. Philos. Mag. Ser.6, **1917**, *34, 200,* 94–98.
42. Perry; R. H.; Green, D. W.; Maloney, J. O.; Perry's Chemical Engineers Handbook, 7th Edition, McGraw-Hill, New York, **1997**, 1–61.
43. Nigmatulin, R. I.; Dynamics of multiphase media.;Moscow, "Nauka," **1987,** *1(2),* 12–14.
44. Kodura, A.; Weinerowska, K.; the influence of the local pipeline leak on water hammer properties, Materials of the II Polish Congress of Environmental Engineering, Lublin, **2005,** 125–133.
45. Kane, J.; Arnett, D.; Remington, B. A.; Glendinning, S. G.; Baz'an G.; "Two-dimensional versus three-dimensional supernova hydrodynamic instability growth,"Astrophys. J.; **2000,** 528–989.
46. Quick, R. S.; "Comparison and Limitations of Various Water hammer Theories,"*J. Hyd. Div. ASME,* May, **1933,** 43–45.
47. Jaeger, C.; "Fluid Transients in Hydro-Electric Engineering Practice," Blackie and Son Ltd.; **1977,** 87–88.
48. Jaime Suárez A.;"Generalized water hammer algorithm for piping systems with unsteady friction" **2005,** 72–77.
49. Fok, A.; Ashamalla, A.; Aldworth, G.; "Considerations in Optimizing Air Chamber for Pumping Plants," Symposium on Fluid Transients and Acoustics in the Power Industry, San Francisco, USA, Dec, **1978,** 112–114.
50. Fok, A.; "Design Charts for Surge Tanks on Pump Discharge Lines," BHRA 3rd Int. Conference on Pressure Surges, Bedford, England, Mar.; **1980,** 23–34.
51. Fok, A.; "Water hammer and Its Protection in Pumping Systems," Hydro technical Conference, CSCE, Edmonton, May,**1982,** 45–55.
52. Fok, A.; "A contribution to the Analysis of Energy Losses in Transient Pipe Flow," PhD.; Thesis, University of Ottawa, **1987,** 176–182.
53. Hariri Asli, K.; Nagiyev, F. B.; Water Hammer and fluid condition, Ministry of Energy, Gilan Water and Wastewater Co.; Research Week Exhibition, Tehran, Iran, December, **2007,** 132–148, http://isrc.nww.co.ir.
54. Hariri Asli, K.; Nagiyev, F. B.; Water Hammer analysis and formulation, Ministry of Energy, Gilan Water and Wastewater Co.; Research Week Exhibition, Tehran, Iran, December, **2007,** 111–131, http://isrc.nww.co.ir.
55. Hariri Asli, K.; Nagiyev, F. B.; Water Hammer and hydrodynamics instabilities, Interpenetration of two fluids at parallel between plates and turbulent moving in pipe, Ministry of Energy, Guilan Water and Wastewater Co.; Research Week Exhibition, Tehran, Iran, December, **2007,** 90–110, http://isrc.nww.co.ir.
56. Hariri Asli, K.; Nagiyev, F. B.; Water Hammer and pump pulsation, Ministry of Energy, Guilan Water and Wastewater Co.; Research Week Exhibition, Tehran, Iran, December, **2007,** 51–72, http://isrc.nww.co.ir.
57. Hariri Asli, K.; Nagiyev, F. B.; Reynolds number and hydrodynamics' instability," Ministry of Energy, Guilan Water and Wastewater Co.; Research Week Exhibition, Tehran, Iran, December, **2007,** 31–50, http://isrc.nww.co.ir.
58. Hariri Asli, K.; Nagiyev, F. B.; Water Hammer and valves, Ministry of Energy, Guilan Water and Wastewater Co.; Research Week Exhibition, Tehran, Iran, December, **2007,** 20–30, http://isrc.nww.co.ir.

59. Hariri Asli, K.; Nagiyev, F. B.; "Interpenetration of two fluids at parallel between plates and turbulent moving in pipe," Ministry of Energy, Guilan Water and Wastewater Co.; Research Week Exhibition, Tehran, Iran, December, 2007, 73–89, http://isrc.nww.co.ir.

60. Hariri Asli, K.; Nagiyev, F. B.; Decreasing of Unaccounted For Water "UFW" by Geographic Information System "GIS" in Rasht urban water system, civil engineering organization of Guilan, Technical and Art Journal, 2007, 3–7, http://www.art-of-music.net/.

61. Hariri Asli, K.; Portable Flow meter Tester Machine Apparatus, Certificate on registration of invention, Tehran, Iran, #010757, Series a/82, 24/11/2007, 1–3.

62. Hariri Asli, K.; Nagiyev, F. B.; Haghi, A. K.; "Interpenetration of two fluids at parallel between plates and turbulent moving in pipe," 9th Conference on Ministry of Energetic works at research week, Tehran, Iran, 2008, 73–89, http://isrc.nww.co.ir.

63. Hariri Asli, K.; Nagiyev, F. B.; Haghi, A. K.; "Water hammer and valves," 9th Conference on Ministry of Energetic works at research week, Tehran, Iran, 2008, 20–30, http://isrc.nww.co.ir.

64. Hariri Asli, K.; Nagiyev, F. B.; Haghi, A. K.; "Water hammer and hydrodynamics instability," 9th Conference on Ministry of Energetic works at research week, Tehran, Iran, 2008, 90–110, http://isrc.nww.co.ir.

65. Hariri Asli, K.; Nagiyev, F. B.; Haghi, A. K.; "Water hammer analysis and formulation," 9th Conference on Ministry of Energetic works at research week, Tehran, Iran, 2008, 27–42, http://isrc.nww.co.ir.

66. Hariri Asli, K.; Nagiyev, F. B.; Haghi, A. K.; "Water hammer &fluid condition," 9th Conference on Ministry of Energetic works at research week, Tehran, Iran, 2008, 27–43, http://isrc.nww.co.ir.

67. Hariri Asli, K.; Nagiyev, F. B.; Haghi, A. K.; "Water hammer and pump pulsation," 9th Conference on Ministry of Energetic works at research week, Tehran, Iran, 2008, 27–44, http://isrc.nww.co.ir.

68. Hariri Asli, K.; Nagiyev, F. B.; Haghi, A. K.; "Reynolds number and hydrodynamics instability," 9th Conference on Ministry of Energetic works at research week, Tehran, Iran, 2008, 27–45, http://isrc.nww.co.ir.

69. Hariri Asli, K.; Nagiyev, F. B.; Haghi, A. K.; "Water hammer and fluid Interpenetration," 9th Conference on Ministry of Energetic works at research week, Tehran, Iran, 2008, 27–47, http://isrc.nww.co.ir.

70. Hariri Asli, K.; GIS and water hammer disaster at earthquake in Rasht water pipeline, civil engineering organization of Guilan, Technical and Art Journal, 2008, 14–17, http://www.art-of-music.net/.

71. Hariri Asli, K.; GIS and water hammer disaster at earthquake in Rasht water pipeline, 3rd International Conference on Integrated Natural Disaster Management, Tehran university, ISSN: 1735–5540, 18–19 Feb.; INDM, Tehran, Iran, 2008, 13, 53/1–12, http://www.civilica.com/Paper-INDM03-INDM03_001.html

72. Hariri Asli, K.; Nagiyev, F. B.; Bubbles characteristics and convective effects in the binary mixtures. Transactions issue mathematics and mechanics series of physical-technical and mathematics science, ISSN: 0002–3108, Azerbaijan, Baku, 2009, 68–74, http://www.imm.science.az/journals.html.

73. Hariri Asli, K.; Nagiyev, F. B.; Haghi, A. K.; Aliyev, S. A.; Three-Dimensional conjugate heat transfer in porous media, 1st Festival on Water and Wastewater Research and Technology, Tehran, Iran, 12–17 Dec. 2009, 26–28, http://isrc.nww.co.ir.

74. Hariri Asli, K.; Nagiyev, F. B.; Haghi, A. K.; Aliyev, S. A.; Some Aspects of Physical and Numerical Modeling of water hammer in pipelines, 1st Festival on Water and Wastewater Research and Technology, Tehran, Iran, 12–17 Dec. 2009, 26–29, http://isrc.nww.co.ir

75. Hariri Asli, K.; Nagiyev, F. B.; Haghi, A. K.; Aliyev, S. A.; Modeling for Water Hammer due to valves: From theory to practice, 1st Festival on Water and Wastewater Research and Technology, Tehran, Iran, 12–17 Dec. 2009, 26, 30, http://isrc.nww.co.ir.

76. Hariri Asli, K.; Nagiyev, F. B.; Haghi, A. K.; Aliyev, S. A.; Water hammer and hydrodynamics instabilities modeling: From Theory to Practice, 1st Festival on Water and Wastewater Research and Technology, Tehran, Iran, 12–17 Dec. 2009, 26–31, http://isrc.nww.co.ir

77. Hariri Asli, K.; Nagiyev, F. B.; Haghi, A. K.; Aliyev, S. A.; A computational approach to study fluid movement, 1st Festival on Water and Wastewater Research and Technology, Tehran, Iran, 12–17 Dec. 2009, 27–32, http://isrc.nww.co.ir.

78. Hariri Asli, K.; Nagiyev, F. B.; Haghi, A. K.; Aliyev, S. A.; Water Hammer Analysis: Some Computational Aspects and practical hints, 1st Festival on Water and Wastewater Research and Technology, Tehran, Iran, 12–17 Dec. 2009, 27–33, http://isrc.nww.co.ir

CHAPTER 6

MATHEMATICAL CONCEPTS AND DYNAMIC MODELING

CONTENTS

6.1 INTRODUCTION

Present book offered numerical modeling for transient flow with approached to engineering aspects and economical hints. On the other hand as a computational approach from theory to practice in numerical analysis modeling, there is computationally efficient for transient flow irreversibility prediction in a practical case. In this book, numerical analysis modeling for reclamation of water transmission line was showed the lining method as the best economical construction way forreclamation.

The greatest development in the theory of water hammer was the analytical methods of calculation. It investigated the hydraulic shock in a simple pipeline (i.e., having a constant diameter and constant speed of propagation of shock waves), by using the general solution of differential equations of unsteady pressure flow. Therefore the equations of water hammer derived in finite differences, which later were called, which were subsequently were used by many researchers in the calculation of water hammer.It was determined by the interaction between the pressure waves that occurred at the pump and reflected in the pipeline. Loss of pressure happened conditionally apart along the pipeline. This method also allows choosing the number and size of shockproof. Development of algorithms for software simulation of transients was made for the complex pressure systems. It included the possible formation of discontinuities flows, hydraulic resistance, structural features of the pumping of water systems (pumps, piping, valves, etc.). However, a calculation of water hammer is adapted to high-pressure water systems for household and drinking purposes. Characteristics method was used for the calculation of water hammer on computer. It was described in detail the process of modeling the unsteady fluid flow in complex piping systems for transporting drinking water.

It was included the description of this phenomenon at discontinuities flow, unsteady friction, changes in gas content and other parameters. Much attention was paid to the way that the original data using a grid, allowing the easiest way to record all raw data for all sites for example, large and extensive network. It was developed a "method of characteristics" for the calculation of water hammer by using computer technology. The process of entering basic information for the calculation of water hammer is simplified for the user by the use of high technology systems. Therefore software can be carried out for multiple calculations of unsteady flow regimes (pressurized systems) in transporting uncontaminated water. Calculations of hydraulic shock in multiphase systems, including a computer, were devoted. Integration of differential equations of unsteady pressure flow is also performed by the "method of characteristics."The method of calculation of water hammer is based on replacing the distributed along the length of the flow of gas parameters concentrated in the fictitious air-hydraulic caps installed on the boundaries of the pipeline. A fictitious elastic element is replaced by elastic deformation of the pipe walls, and the elastic deformation of the solid suspension is modeled by fictitious elastic elements of the solid suspension. However, detailed experimental studies are based on the solid component. A particular challenge in terms of calculations is a hydraulic shock, accompanied by discontinuities of the flow.

This phenomenon has not been fully studied, so the works of many scientists are devoted to experimental and theoretical studies of nonstationary processes with a break of the column of liquid. Researches devoted to the study of water hammer to break the continuity of flow in the intermediate point.It analytically derived the condition for the gap of continuous flow at a turning point of the pipeline when the pressure is reduced at this point (below atmospheric pressure). Studies have shown that the location of the discontinuity of continuous flow at a turning point depends, first of all, the profile of the pipeline. Protection of hydraulic systems against water hammer by releasing part of the transported fluid is the most widespread method of artificial reduction of the hydraulic shock. Devices that perform this function can be divided into valve, bursting disc and the overflow of the column [1-17].

6.2 MATERIALS AND METHODS

This work presented a suitable way for economical analyzing of construction related to transient flow destruction. Transient flow is solved for the pipeline in the range of approximate equations. These approximate equations are solved by numerical solutions of the nonlinear Navier–Stokes equations in a method of characteristics "MOC." So, experiences have been ensured for the reliable water transmission pipeline. Numerical modeling and simulation which was defined by method of characteristics "MOC" have been provided a set of results.

The method of characteristics "MOC" approach transforms the water hammer partial differential equations into the ordinary differential equations along the characteristic lines defined as the continuity equation and the momentum equation are needed to determine V and P in a one dimensional flow system. Solving these two equations produces a theoretical result that usually corresponds quite closely to actual system measurements if the data and assumptions used to build the numerical model are valid [18-33].

6.2.1 APPARATUS

Transient analysis results that are not comparable with actual system measurements are generally caused by inappropriate system data (especially boundary conditions) and inappropriate assumptions. Comparisons between the models and validation data can be grouped into the following three categories: (a) Cases for which closed-form analytical solutions exist given certain assumptions if the model can directly reproduce the solution, is considered valid for this case. (b) Laboratory experiments with flow and pressure data records. The model is calibrated using one set of data and, without changing parameter values, it is used to match a different set of results. If successful, it is considered valid for these cases. (c) Field tests on actual systems with flow and pressure data records. These comparisons require threshold and span calibration of all sensor groups, multiple simultaneous datum, and time base checks and careful test planning and interpretation. Sound calibrations match multiple sensor records and reproduce both peak timing and secondary signals all measured every second or fraction of a second [34-43].

A model for liquid- vapor flows illustrates the numerical techniques for solving the resulting equations. Hence field test model was chosen for experimental presentation ofwater hammer phenomenon at the water pipeline [1–18].

$$\left(P_1 / \gamma\right) + Z_1 + \left(V_1^2 / 2g\right) + h_p = \left(P_2 / \gamma\right) + Z_2 + \left(V_2^2 / 2g\right) + h_L, \tag{1}$$

$$\left(g / a\right)\left(dH / dt\right) + dv / dt + \left(f\, v|v|2d\right) = 0 \Rightarrow \left(ds / dt\right) = c^+, \tag{2}$$

$$-\left(g / a\right)\left(dH / dt\right) + dv / dt + \left(f\, v|v|2d\right) = 0 \Rightarrow \left(ds / dt\right) = c^-, \tag{3}$$

The method of characteristics is a finite difference technique where pressures were computed along the pipe for each time step. Calculation automatically subdivided the pipe into sections (intervals) and selected a time interval for computations.

$$\left(dp / dt\right) = \left(\partial p / \partial t\right) + \left(\partial p / \partial s\right)\left(ds / dt\right), \tag{4}$$

$$(dv \,/\, dt) = (\partial vv \,/\, \partial t) + (\partial v \,/\, \partial s)(ds \,/\, dt), \qquad (5)$$

P and V changes due to time are high and due to coordination are low then it can be neglected for coordination differentiation:

$$(\partial v \,/\, \partial t) + (1 \,/\, \rho)(\partial \rho \,/\, \partial s) + g(dz \,/\, ds) + (f \,/\, 2D)v|v| = 0,$$

$$\text{(Euler equation), (6)}$$

$$C^2(\partial v \,/\, \partial s) + (1 \,/\, P)(\partial P \,/\, \partial t) = 0, \quad \text{(Continuity equation),} \qquad (7)$$

By linear combination of Euler and continuity equations in characteristic solution Method:

$$\lambda\big[(\partial v \,/\, \partial t) + (1 \,/\, \rho)(\partial p \,/\, \partial s) + g(dz \,/\, ds) + (f \,/\, 2D)v|v|\big] + C^2(\partial v \,/\, \partial s) + (1 \,/\, p)(\partial p \,/\, \partial t) = 0,$$
$$\lambda =^+ c \,\&\, \lambda =^- c$$

$$(8)$$

$$(dv \,/\, dt) + (1 \,/\, cp)(dp \,/\, ds) + g(dz \,/\, ds) + (f \,/\, 2D)v|v| = 0, \qquad (9)$$

$$(dv \,/\, dt) - (1 \,/\, cp)(\partial p \,/\, \partial s) + g(dz \,/\, ds) + (f \,/\, 2D)v|v| = 0, \qquad (10)$$

Method of characteristics drawing in $(s\text{--}t)$ coordination:

$$(dv \,/\, dt) - (g \,/\, c)(dH \,/\, dt) = 0, \qquad (11)$$

$$dH = (c \,/\, g)dv, \text{ (Joukowski Formula),} \qquad (12)$$

By Finite Difference method:

$$c+:\left((vp-v_{Le})(Tp-0)\right)+\left((g/c)(Hp-H_{Le})/(Tp-0)\right)+\left((fv_{L_e}|v_{L_e}|)/2D\right)=0\Big|, \quad (13)$$

$$c-:\left((vp-vRi)(Tp-0)\right)+\left((g/c)(Hp-HRi)/(Tp-0)\right)+\left((fvRi|vRi|)/2D\right)=0\Big|, \quad (14)$$

$$c+:\left(vp-v_{Le}\right)+\left(g/c\right)\left(Hp-H_{Le}\right)+\left(f\Delta t\right)\left(fv_{Le}|v_{Le}|\right)/2D=0, \quad (15)$$

$$c-:\left(vp-vRi\right)+\left(g/c\right)\left(Hp-HRi\right)+\left(f\Delta t\right)\left(v_{Ri}|v_{Ri}|\right)/2D=0 \quad , (16)$$

$$V_p=1/2\left(\frac{\left(V_{Le}+V_{ri}\right)+\left(g/c\right)\left(H_{Le}-H_{ri}\right)}{-\left(f\,\Delta t/2D\right)\left(V_{Le}\,|V_{Le}|+V_{ri}|V_{ri}|\right)}\right), \quad (17)$$

$$H_p=1/2\left(\frac{C/g\left(V_{Le}-V_{ri}\right)+\left(H_{Le}+H_{ri}\right)}{-C/g\left(f\,\Delta t/2D\right)\left(V_{Le}\,|V_{Le}|-V_{ri}|V_{ri}|\right)}\right), \quad (18)$$

6.3   RESULTS

6.3.1   HIGH-PRESSURE DROP INWATER PIPELINE

For make decision about reclamation of leakage in water transmission line (Table1).

TABLE 1 Comparison of simulation and measurement of high-speed sensors and equipment (field trials) observed during the hydraulic shock.

Pressure	Flow	Distance	Time	Pressure
(m-Hd)	(lit/s)	(m)	(s)	(m-Hd)
Field				Model
86	2491	3390	0	89.033
86	2491	3390	1	89.764
88	2520	3291	0	90.427
90	2520	3190	1	91.663
95	2574	3110	1.4	94.03
95	2574	3110	1.4	94.03
95	2574	3110	1.5	94.1
95	2590	3110	2	94.96
95	2590	3110	2	94.96
95.7	2600	3110	2	95.27
95.7	2600	3110	3	96.73
95.7	2600	3110	4	96.73
95.7	2600	3110	5	97.46
95.7	2605	3110	0.5	94.33
100	2633	2184	1.3	100.41
100	2633	2928	1.3	96.69
101	2650	2920	1.4	97.33
106	2680	1483	1.4	105.45
107	2690	1217	1.4	107.09
109	2710	1096	1.4	108.31
109	2710	1096	1.4	108.31
110	2920	1000	1.5	115.37

The water hammer modeling discuses for three cases:

- Simulation of pipeline (Fig. 1) for reclamation numerical analysis of water transmission line by replace of present reinforced concrete pipe (AC pipe) with the same diameter of polyethylene pipe (AC pipe-1200 mm replace withPE pipe-1200 mm).
- Simulation of pipeline (Fig. 2) for reclamation numerical analysis of water transmission line by lining (push a new smaller diameter pipe into the old larger diameter pipe-Push PE pipe-1100 mm into AC pipe-1200 mm).
- Simulation of pipeline (Fig. 2) for reclamation numerical analysis of water transmission line by replace of present reinforced concrete pipe (AC pipe) with the larger diameter of polyethylene pipe (AC pipe-1200 mm replace with PE pipe-1300 mm).

According to pipeline specification three models were defined.The leakage point as the most important critical point was selected for analysis byabove three models for pipeline reclamation numerical analysis. Comparison of these three cases revealed the reclamation numerical analysis curves as long as water transmission line at transient flow condition[45-63].

For all three models were showed the numerical analysis of existent pipe reclamation (reinforced concrete pipe replacement with the offered polyethylene pipe) as flowing procedure:

- *Max.* Pressure–Distance; *Min.* Pressure–Distance curves Pressure drop is high proportional to diameter increasing.
- *Max.* Pressure–Distance curve shows pressure rising. Pressure rising became low and low proportional to diameter increasing.
- *Min.* Pressure–Distance curve shows pressure decreasing. So pressure drop became low and low proportional to diameter increasing.
- With polyethylene pipe diameter increasing, *Min.* Pressure and Max. Pressure drop happened at the near of pump station location.
  Max. Pressure drop happened when diameter changed from small diameter to large diameter.
  Min. Pressure drop happened when diameter changed from large diameter to small diameter.
- Pressure variation was decreased by diameter increasing [64-73].

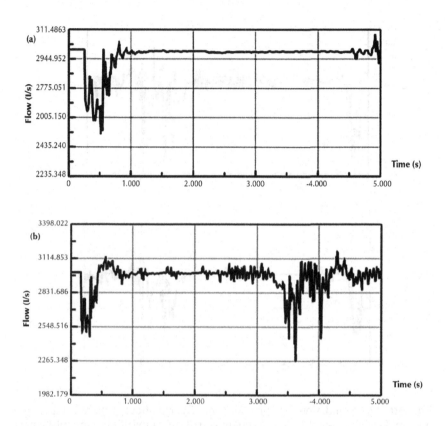

FIGURE 1 Simulation of pipeline for Reclamation: (a) AC pipe-1200 mm, (b) AC pipe-1200 mm replacement with PE pipe-1200 mm.

Reclamation numerical analysis modeling showed the best construction way for water transmission line was the lining of present reinforced concrete pipe (AC pipe). This variant for reclamation was based on lining of present reinforced concrete pipe with the smaller diameter of polyethylene pipe (AC pipe-1200 mm must be replaced by PE pipe-1100 mm). It was the best construction way for reclamation. But the reclamation numerical modeling showed pressure drop happened when diameter changed from large diameter to smaller diameter. Many factors including: Total budgets, Time of project and etc., were redoundto selection of the two variants: (a) Reclamation of water transmission line by lining, (b) Reclamation of water transmission line by replacement of existent pipe with the larger diameter of polyethylene pipe (AC pipe-1200 mm replace withPE pipe-1300 mm).

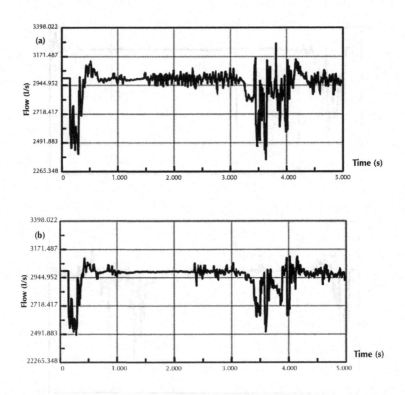

FIGURE 2 Simulation of pipeline for Reclamation: (a) AC pipe-1200 mm replacement withPE pipe-1100 mm, (b) AC pipe-1200 mmreplacement withPE pipe-1300 mm.

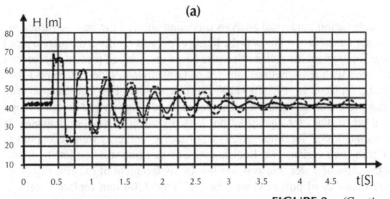

FIGURE 3 *(Continued)*

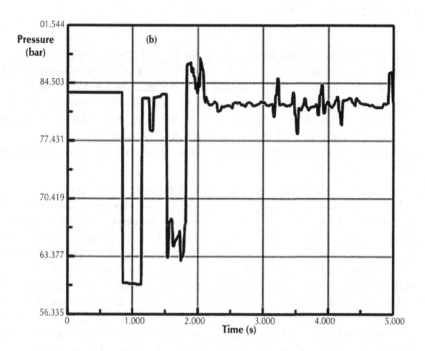

FIGURE 3 Experimental observed and calculated results for (a) Kodura and Weinerowska research, (b) present research.

6.3.2 COMPARISON OF PRESENT RESEARCH RESULTS WITH OTHER EXPERT'S RESEARCH

Apoloniusz Kodura, and Katarzyna Weinerowska [4], In the work of Kodura and Weinerowska water hammer in pipeline at the local leak case have been presented (Fig. 3). This case was related to some additional factors. Therefore detailed conclusions drawn on the basis of experiments and calculations for the pipeline were presented in the paper of Kodura and Weinerowska [4]. The most important points, which were observed are as flowing: These effects were studied related to the values for period of oscillations. In a consequence it was studied related to the value of wave celerity when the outflow to the overpressure reservoir [74-84].

6.4 CONCLUSIONS

Transient analysis should be performed for large, high-value pipelines, especially with pump stations.A complete transient analysis, in conjunction with other system design activities, should be performed during the initial design phases of a project. Normal flow-control operations and predicable emergency operations should, was evaluated during the work. Make decision process from theory to practice for reclamation of water transmission line was considered in this work. It confirmed by nonlinear heterogeneous model for water hammer in three cases.

Reclamation numerical modeling showed pressure decreasing became high and high proportional to diameter increasing. It also showed pressure decreasing became low and low proportional to diameter increasing. *Max.* Pressure drop happened when diameter changed from smaller diameter to larger diameter and *Min.* Pressure decreasing happened when diameter changed from larger diameter to smaller diameter. Pressure rising became low and low proportional to diameter increasing. *Max.* pressure drop happened as long as diameter changed from small diameter to larger diameter and *Min.* pressure drop happened when diameter changed from large diameter to smaller diameter. This was showed the numerical analysis modelingas a computational approach is computationally efficient for transient flow irreversibility prediction in a practical case. It offered the lining method as a construction way forreclamation of damaged water transmission line. The reason for offering of this variant for reclamation was based on the lining of present reinforced concrete pipe with smaller diameter of polyethylene pipe (AC pipe-1200 mm must be replaced by PE pipe-1100 mm).

KEYWORDS

- **Economical hints**
- **Irreversibility**
- **Transient flow**
- **Water hammer**

REFERENCES

1. Arturo Leon, S. improved modeling of unsteady free surface, pressurized and mixed flows in storm-sewer systems, Submitted in partial fulfillment of the requirements for the degree of Doctor of Philosophy in Civil Engineering in the Graduate College of the University of Illinois at Urbana-Champaign, 2007, 57–58.
2. Hariri Asli, K.; Nagiyev, F. B.; Haghi, A. K. Computational methods in applied science and engineering. In: Interpenetration of Two Fluids at Parallel Between Plates and Turbulent Moving in Pipe. Nova Science, New York, USA, 2009, 115–128 Chapter7, https://www.novapublishers.com/catalog/ product_ info.php?products_id=10681.
3. Wylie, E. B.; Streeter, V. L. Fluid transients, Feb Press, Ann. Arbor, MI, 1983, corrected copy: 1982, 166–171.
4. Apoloniusz, Kodura, Katarzyna, Weinerowska.: Some aspects of physical and numerical modeling of Water Hammer in Pipelines, 2005, 125–133.
5. Hariri Asli, K.; Nagiyev, F. B.; Haghi A. K. "Some Aspects of Physical and Numerical Modeling of water hammer in pipelines," Nonlinear DynamicsAn International Journal of Nonlinear Dynamics and Chaos in Engineering Systems, ISSN: 1573–269X (electronic version) Journal no. 11071 Springer, Netherland, ISSN: 0924–090X (Print version), Germany, 677–701, Volume 60, Number 4 / June, 2010. http://www.springerlink.com/openurl.asp?genre=article&id=doi: 10.1007/s11071-009-9624-7.
6. Hariri Asli, K.; Nagiyev, F. B.; Haghi, A. K.; Aliyev S. A. Physical and Numerical Modeling of Fluid Flow in Pipelines: A computational approach, International Journal of the Balkan Tribological Association, ISSN: 1310–4772, Sofia, Bulgaria, 2010, 16(1), 20–34.
7. Hariri Asli, K.; Nagiyev, F. B.; Haghi, A. K.; Aliyev S. A. Nonlinear Heterogeneous Model for Water Hammer Disaster, International Journal of the Balkan Tribological Association, ISSN: 1310–4772, Sofia, Bulgaria, 2010, 16, 2, 209–222.
8. Hariri Asli, K.; Nagiyev, F. B.; Haghi, A. K.; Aliyev S. A, Hariri Asli H. Numerical modeling of transients flow in water pipeline: A computational approach, International Journal of Academic Research, ISSN: 1310–4772, Baku, Azerbaijan, ISSN: 2075–4124, 2010, 2(5), September 30.
9. Lee, T. S.; Pejovic, S. Air influence on similarity of hydraulic transients and vibrations, ASME J. Fluid Eng. 1996, 118(4), 706–709.
10. Fedorov, A. G.; Viskanta R.; Three-dimensional Conjugate Heat Transfer into Microchannel Heat Sink for Electronic Packaging, Int. J.; Heat Mass Transfer2000, 43, 399–415.
11. Tuckerman, D. B.; Heat transfer microstructures for integrated circuits, Ph.D. thesis, Stanford University, 1984, 10–120.
12. Harms, T. M.; Kazmierczak, M. J.; Cerner, F. M.; Holke A.; Henderson, H. T.; Pilchowski, H. T.; Baker K.; Experimental Investigation of Heat Transfer and Pressure Drop through Deep Micro channels in a (100) Silicon Substrate, in: Proceedings of the ASME.; Heat Transfer Division, HTD 1997, 351, 347–357.
13. Holland, F. A.; Bragg R.; Fluid Flow for Chemical Engineers, Edward Arnold Publishers, London, 1995, 1–3.

14. Lee, T. S.; Pejovic, S. Air influence on similarity of hydraulic transients and vibrations. *ASME J. Fluid Eng.* **1996,** *118(4),* 706–709.
15. Li J.; McCorquodale A.; "Modeling Mixed Flow in Storm Sewers," Journal of Hydraulic Engineering, ASCE, **1999,** *125(11),* 1170–1180.
16. Minnaert M.; on musical air bubbles and the sounds of running water. *Phil. Mag.* **1933,** v. *16(7),* 235–248.
17. Moeng, C. H.; McWilliams, J. C.; Rotunno, R.; Sullivan, P. P.; Weil, J.; "Investigating 2D modeling of atmospheric convection in the PBL,"*J. Atm. Sci.* **2004,** *61,* 889–903.
18. Tuckerman, D. B.; R. F. W Pease, high performance heat sinking for VLSI, IEEE Electron device letter, DEL-2, **1981,** 126–129.
19. Nagiyev, F. B.; Khabeev N. S. Bubble dynamics of binary solutions. High Temperature, **1988,** *27(3),* 528–533.
20. Shvarts D.; Oron D.; Kartoon D.; Rikanati A.; Sadot O.; "Scaling laws of non-linear Rayleigh-Taylor and Richtmyer-Meshkov instabilities in two and three dimensions,"*C.R. Acad. Sci. Paris IV,* **2000,** *719,* 312 p.
21. Cabot, W. H.; Cook, A. W.; Miller, P. L.; Laney, D. E.; Miller, M. C.; Childs, H. R.; "Large eddy simulation of Rayleigh-Taylor instability," Phys. Fluids, September, **2005,** *17,* 91–106.
22. Cabot W.; University of California, Lawrence Livermore National laboratory, Livermore, CA, *Phys. Fluids,* **2006,** 94–550.
23. Goncharov V. N.; "Analytical model of nonlinear, single-mode, classical Rayleigh-Taylor instability at arbitrary Atwood numbers,"*Phys. Rev. Lett.* **2002,** *88,* 134502, 10–15.
24. Ramaprabhu, P.; Andrews, M. J.; "Experimental investigation of Rayleigh-Taylor mixing at small Atwood numbers,"*J. Fluid Mech.* **2004,** *502,* 233 p.
25. Clark, T. T.; "A numerical study of the statistics of a two-dimensional Rayleigh-Taylor mixing layer,"*Phys. Fluids* **2003,** *15,* 2413.
26. Cook, A. W.; Cabot W.; Miller, P. L.; "The mixing transition in Rayleigh-Taylor instability,"*J. Fluid Mech.* **2004,** *511,* 333.
27. Waddell, J. T.; Niederhaus, C. E.; Jacobs, J. W.; "Experimental study of Rayleigh-Taylor instability: Low Atwood number liquid systems with single-mode initial perturbations,"*Phys. Fluids* **2001,** *13,* 1263–1273.
28. Weber, S. V.; Dimonte, G.; Marinak, M. M.; "Arbitrary Lagrange-Eulerian code simulations of turbulent Rayleigh-Taylor instability in two and three dimensions," Laser and Particle Beams**2003,** *21,* 455 p.
29. Dimonte G.; Youngs D.; Dimits A.; Weber S.; Marinak M. "A comparative study of the Rayleigh-Taylor instability using high-resolution three-dimensional numerical simulations: the Alpha group collaboration,"*Phys. Fluids* **2004,** *16,* 1668.
30. Young, Y. N.; Tufo H.; Dubey A.; Rosner R.; "On the miscible Rayleigh-Taylor instability: two and three dimensions,"*J. Fluid Mech.***2001,** *447, 377,* 2003–2500.
31. George E.; Glimm J.; "Self-similarity of Rayleigh-Taylor mixing rates,"*Phys. Fluids* **2005,** *17,* 054101, 1–3.
32. Oron. D.; Arazi L.; Kartoon D.; Rikanati A.; Alon U.; Shvarts D.; "Dimensionality dependence of the Rayleigh-Taylor and Richtmyer-Meshkov instability late-time scaling laws,"*Phys. Plasmas* **2001,** *8,* 2883.

33. Nigmatulin, R. I.; Nagiyev, F. B.; Khabeev, N. S.; Effective heat transfer coefficients of the bubbles in the liquid radial pulse. Mater. Second-Union. Conf. Heat Mass Transfer, "Heat massoob-men in the biphasic with Minsk,"**1980,** *5,* 111–115.

34. Nagiyev, F. B.; Khabeev N. S, Bubble dynamics of binary solutions. High Temperature, **1988,** v. *27(3),* 528–533.

35. Nagiyev, F. B.; Damping of the oscillations of bubbles boiling binary solutions. Mater. VIII Resp. Conf. mathematics and mechanics. Baku, October 26–29, **1988,** 177–178.

36. Nagyiev, F. B.; Kadyrov, B. A.; Small oscillations of the bubbles in a binary mixture in the acoustic field Math. An Az. SSR Ser. Physicotech. and mate. Science, **1986,** *1,* 23–26.

37. Nagiyev, F. B.; Dynamics, heat and mass transfer of vapor-gas bubbles in a two-component liquid. Turkey-Azerbaijan petrol semin.; Ankara, Turkey, **1993,** 32–40.

38. Nagiyev, F. B.; The method of creation effective coolness liquids, Third Baku international Congress. Baku, Azerbaijan Republic, **1995,** 19–22.

39. Nagiyev, F. B.; The linear theory of disturbances in binary liquids bubble solution. Dep. In VINITI, **1986,** *405, 86,* 76–79.

40. Nagiyev, F. B.; Structure of stationary shock waves in boiling binary solutions. Math. USSR, Fluid Dynamics, **1989,** *1,* 81–87.

41. Rayleigh, On the pressure developed in a liquid during the collapse of a spherical cavity. Philos. Mag. Ser.6, **1917,** *34, 200,* 94–98.

42. Perry.; R. H.; Green, D. W.; Maloney, J. O.; Perry's Chemical Engineers Handbook, 7th Edition, McGraw-Hill, New York, **1997,** 1–61.

43. Nigmatulin, R. I.; Dynamics of multiphase media. Moscow, "Nauka," **1987,** *1(2),* 12–14.

44. Kodura, A.; Weinerowska, K.; the influence of the local pipeline leak on water hammer properties, Materials of the II Polish Congress of Environmental Engineering, Lublin, **2005,** 125–133.

45. Kane J.; Arnett D.; Remington, B. A.; Glendinning, S. G.; Baz'an G.; "Two-dimensional versus three-dimensional supernova hydrodynamic instability growth," Astrophys. J.; **2000,** 528–989.

46. Quick, R. S.; "Comparison and Limitations of Various Water hammer Theories,"*J. Hyd. Div. ASME,* May, **1933,** 43–45.

47. Jaeger C.; "Fluid Transients in Hydro-Electric Engineering Practice," Blackie and Son Ltd.; **1977,** 87–88.

48. Jaime Suárez A.; "Generalized water hammer algorithm for piping systems with unsteady friction" **2005,** 72–77.

49. Fok, A.; Ashamalla, A.; Aldworth, G.; "Considerations in Optimizing Air Chamber for Pumping Plants," Symposium on Fluid Transients and Acoustics in the Power Industry, San Francisco, USA, Dec, **1978,** 112–114.

50. Fok, A.; "Design Charts for Surge Tanks on Pump Discharge Lines," BHRA 3rd Int. Conference on Pressure Surges, Bedford, England, Mar.; **1980,** 23–34.

51. Fok, A.; "Water hammer and Its Protection in Pumping Systems," Hydro technical Conference, CSCE, Edmonton, May, **1982,** 45–55.

52. Fok, A.; "A contribution to the Analysis of Energy Losses in Transient Pipe Flow," PhD.; Thesis, University of Ottawa, **1987,** 176–182.

53. Hariri Asli, K.; Nagiyev, F. B.; Water Hammer and fluid condition, Ministry of Energy, Gilan Water and Wastewater Co.; Research Week Exhibition, Tehran, Iran, December, **2007,** 132–148, http://isrc.nww.co.ir.

54. Hariri Asli, K.; Nagiyev, F. B.; Water Hammer analysis and formulation, Ministry of Energy, Gilan Water and Wastewater Co.; Research Week Exhibition, Tehran, Iran, December, **2007,** 111–131, http://isrc.nww.co.ir.

55. Hariri Asli, K.; Nagiyev, F. B.; Water Hammer and hydrodynamics instabilities, Interpenetration of two fluids at parallel between plates and turbulent moving in pipe, Ministry of Energy, Guilan Water and Wastewater Co.; Research Week Exhibition, Tehran, Iran, December, **2007,** 90–110, http://isrc.nww.co.ir.

56. Hariri Asli, K.; Nagiyev, F. B.; Water Hammer and pump pulsation, Ministry of Energy, Guilan Water and Wastewater Co.; Research Week Exhibition, Tehran, Iran, December, **2007,** 51–72, http://isrc.nww.co.ir.

57. Hariri Asli, K.; Nagiyev, F. B.; Reynolds number and hydrodynamics' instability," Ministry of Energy, Guilan Water and Wastewater Co.; Research Week Exhibition, Tehran, Iran, December, **2007,** 31–50, http://isrc.nww.co.ir.

58. Hariri Asli, K.; Nagiyev, F. B.; Water Hammer and valves, Ministry of Energy, Guilan Water and Wastewater Co.; Research Week Exhibition, Tehran, Iran, December, **2007,** 20–30, http://isrc.nww.co.ir.

59. Hariri Asli, K.; Nagiyev, F. B.; "Interpenetration of two fluids at parallel between plates and turbulent moving in pipe," Ministry of Energy, Guilan Water and Wastewater Co.; Research Week Exhibition, Tehran, Iran, December, **2007,** 73–89, http://isrc.nww.co.ir.

60. Hariri Asli, K.; Nagiyev, F. B.; Decreasing of Unaccounted For Water "UFW" by Geographic Information System"GIS" in Rasht urban water system, civil engineering organization of Guilan, Technical and Art Journal, **2007,** 3–7, http://www.art-of-music.net/.

61. Hariri Asli, K.; Portable Flow meter Tester Machine Apparatus, Certificate on registration of invention, Tehran, Iran, #010757, Series a/82, 24/11/2007, 1–3

62. Hariri Asli, K.; Nagiyev, F. B.; Haghi, A. K.; "Interpenetration of two fluids at parallel between plates and turbulent moving in pipe," 9th Conference on Ministry of Energetic works at research week, Tehran, Iran, **2008,** 73–89, http://isrc.nww.co.ir.

63. Hariri Asli, K.; Nagiyev, F. B.; Haghi, A. K.; "Water hammer and valves," 9th Conference on Ministry of Energetic works at research week, Tehran, Iran, **2008,** 20–30, http://isrc.nww.co.ir.

64. Hariri Asli, K.; Nagiyev, F. B.; Haghi, A. K.; "Water hammer and hydrodynamics instability," 9th Conference on Ministry of Energetic works at research week, Tehran, Iran, **2008,** 90–110, http://isrc.nww.co.ir.

65. Hariri Asli, K.; Nagiyev, F. B.; Haghi, A. K.; "Water hammer analysis and formulation," 9th Conference on Ministry of Energetic works at research week, Tehran, Iran, **2008,** 27–42, http://isrc.nww.co.ir.

66. Hariri Asli, K.; Nagiyev, F. B.; Haghi, A. K.; "Water hammer &fluid condition," 9th Conference on Ministry of Energetic works at research week, Tehran, Iran, **2008,** 27–43, http://isrc.nww.co.ir.

67. Hariri Asli, K.; Nagiyev, F. B.; Haghi, A. K.; "Water hammer and pump pulsation," 9th Conference on Ministry of Energetic works at research week, Tehran, Iran, **2008**, 27–44, http://isrc.nww.co.ir.
68. Hariri Asli, K.; Nagiyev, F. B.; Haghi, A. K.; "Reynolds number and hydrodynamics instability," 9th Conference on Ministry of Energetic works at research week, Tehran, Iran, **2008**, 27–45, http://isrc.nww.co.ir.
69. Hariri Asli, K.; Nagiyev, F. B.; Haghi, A. K.; "Water hammer and fluid Interpenetration," 9th Conference on Ministry of Energetic works at research week, Tehran, Iran, **2008**, 27–47, http://isrc.nww.co.ir.
70. Hariri Asli, K.; GIS and water hammer disaster at earthquake in Rasht water pipeline, civil engineering organization of Guilan, Technical and Art Journal, **2008**, 14–17, http://www.art-of-music.net/.
71. Hariri Asli, K.; GIS and water hammer disaster at earthquake in Rasht water pipeline, 3rd International Conference on Integrated Natural Disaster Management, Tehran university, ISSN: 1735–5540, 18–19 Feb.; INDM, Tehran, Iran, **2008**, *13*, 53/1–12, http://www.civilica.com/Paper-INDM03-INDM03_001.html
72. Hariri Asli, K.; Nagiyev, F. B.; Bubbles characteristics and convective effects in the binary mixtures. Transactions issue mathematics and mechanics series of physical-technical and mathematics science, ISSN: 0002–3108, Azerbaijan, Baku, **2009**, 68–74, http://www.imm.science.az/journals.html.
73. Hariri Asli, K.; Nagiyev, F. B.; Haghi, A. K.; Aliyev, S. A.; Three-Dimensional conjugate heat transfer in porous media, 1st Festival on Water and Wastewater Research and Technology, Tehran, Iran, 12–17 Dec. **2009**, 26–28, http://isrc.nww.co.ir.
74. Hariri Asli, K.; Nagiyev, F. B.; Haghi, A. K.; Aliyev, S. A.; Some Aspects of Physical and Numerical Modeling of water hammer in pipelines, 1st Festival on Water and Wastewater Research and Technology, Tehran, Iran, 12–17 Dec. **2009**, 26–29, http://isrc.nww.co.ir
75. Hariri Asli, K.; Nagiyev, F. B.; Haghi, A. K.; Aliyev, S. A.; Modeling for Water Hammer due to valves: From theory to practice, 1st Festival on Water and Wastewater Research and Technology, Tehran, Iran, 12–17 Dec. **2009**, 26, 30, http://isrc.nww.co.ir.
76. Hariri Asli, K.; Nagiyev, F. B.; Haghi, A. K.; Aliyev, S. A.; Water hammer and hydrodynamics instabilities modeling: From Theory to Practice, 1st Festival on Water and Wastewater Research and Technology, Tehran, Iran, 12–17 Dec. **2009**, 26–31, http://isrc.nww.co.ir
77. Hariri Asli, K.; Nagiyev, F. B.; Haghi, A. K.; Aliyev, S. A.; A computational approach to study fluid movement, 1st Festival on Water and Wastewater Research and Technology, Tehran, Iran, 12–17 Dec. **2009**, 27–32, http://isrc.nww.co.ir.
78. Hariri Asli, K.; Nagiyev, F. B.; Haghi, A. K.; Aliyev, S. A.; Water Hammer Analysis: Some Computational Aspects and practical hints, 1st Festival on Water and Wastewater Research and Technology, Tehran, Iran, 12–17 Dec. **2009**, 27–33, http://isrc.nww.co.ir 1. Leon, S. A.; Improved Modeling of Unsteady Free Surface, Pressurized and Mixed Flows in Storm-Sewer Systems, Submitted in Partial Fulfillment of the Requirements for the degree of Doctor of Philosophy in Civil Engineering in the Graduate College of the University of Illinois at Urbana-Champaign, **2007**, 57–58.

79. Hariri Asli, K.; Nagiyev F. B.; Haghi, A. K.; Physical and Numerical Modeling of Fluid Flow in Pipelines, A computational approach.*Int. J. Balkan Tribological Association,* Thomson Reuters Master Journal List, ISSN: 1310–4772. Sofia, Bulgaria, 16. **2009,** *19,* 20–34.

80. Hariri Asli, K.; Nagiyev, F. B.; Haghi, A. K.; Interpenetration of Two Fluids at Parallel between Plates and Turbulent Moving in Pipe, Computational Methods in Applied Science and Engineering, Nova Science Publications, New York, USA, chapter7, **2010,** 115–128.

81. Wylie, E. B.V. L.; Streeter (7): Fluid Transients in Systems. Prentice Hall, **1993,** corrected copy, **1982,** 166–171.

82. Kodura, A.; Weinerowska K.; Some Aspects of Physical and Numerical Modeling of Water Hammer in Pipelines Ottenstein, Austria, **2005,** 125–133.

83. Hariri Asli, K.; Nagiyev F. B.; Haghi, A. K.; Some Aspects of Physical and Numerical Modeling of Water Hammer in Pipelines, Nonlinear Dynamics. International *J.* Nonlinear Dynamics and Chaos in Engineering Systems, ISSN: 0924–090X, (electronic version), Journal 11071 Springer, Published online: 10 December, **2009,** (print version), ISSN: 1573–269X 60. **2010,** *4,* 677–701.

84. Hariri Asli, K.; Nagiyev, F. B.; Haghi, A. K.; Water hammer analysis; some computational aspects and practical hints, Computational Methods in Applied Science and Engineering, Nova Science Publications, New York, USA, chapter16, **2010,** 263–282.

CHAPTER 7

MODELING FOR PREDICTIONS OF AIR ENTRANCE INTO WATER PIPELINE

CONTENTS

7.1   INTRODUCTION

One of the most important factors for water hammer solution is formed by predicted the air entrance or, rate and location of air penetrated into the water pipeline. Water hammer can be recognized during irregular operations. Under abnormal operations (during an earthquake) at power turning on and off, water hammer will attack to the system. The equipment, which can help during an emergency outage, includes surge tank, switching tanks, boilers, valves, pressure vessel for the pulse pressure damping, and pressure relief valves.

It is necessary to background information in the database. It needs to the exchange the data between the receiver and transmitter. The pipeline was equipped with the Program Logic Control (PLC). It can be controlled the entire system in online mode, by sending voltage to the valves, pumps, protective shells from PLC. In this way it can protect the system from the water hammer disaster. In some cases, hydraulic force in the transitional flow regimes lead to cracks or ruptures of pipes, even at low speeds and steady flow.

It will also explore the influence of system topology, the characteristics of fluid on the most likely causes for transients. N. E. Zhukovsky introduced the concept of the effective sound speed. He mentioned to reducing the motion of a compressible fluid in an elastic cylindrical pipe to the motion of a compressible fluid in a rigid pipe, but with a lower modulus of elasticity of the liquid. Calculations of hydraulic shock in multiphase systems, including a computer, are devoted to the work of V. M. Alysheva. In that work, integration of differential equations of unsteady pressure flow is also performed by the "method of characteristics." The works of Streeter, K. P. Vishnevsky, B. F. Lyamaeva, and V. M. Alyshev use the method of calculation of water hammer. They are based on replacing the distributed along the length of the flow of gas parameters concentrated in the fictitious air-hydraulic caps installed on the boundaries of the pipeline. A fictitious elastic element is replaced by elastic deformation of the pipe walls, and the elastic deformation of the solid suspension is modeled by fictitious elastic elements of the solid suspension. However, detailed experimental studies are based on the solid component. First detailed study and writing the first design formula, for such cases of hydraulic shocks with discontinuities of continuous flow, was the work of A. F. Masty [1-11].

7.2   MATERIAL AND METHODS

The pressure wave speed is a fundamental parameter for hydraulic transient modeling at present research, since it determined how quickly disturbances propagate throughout the system. This affected whether or not different pulses may superpose or cancel each other as they meet at different times and locations. Wave speed was affected by pipe material and bedding, as well as by the presence of fine air bubbles in the fluid. The default value of 1,000 (m/s), 3, 280 (ft/s) was for metal or concrete pipe.

For present research with free gas systems (Figs. 1-19) and the potential for water-column separation, the numerical simulation of hydraulic transients was more complex. Small pressure spikes caused by the type of tiny vapor pockets that was difficult to simulate accurately seldom result in a significant change to the transient envelopes. Larger vapor-pocket collapse events resulting in significant upsurge pressures were simulated with enough accuracy to support definitive conclusions.

With respect to timing, there were close agreement between the Computed and measured periods of the system, regardless of what flow-control operation initiated the transient. With a well-calibrated model of the system, it was possible to use the model in the operational control of the system and anticipated the effects of specific flow-control operations. This required field measurements to quantify system's pressure wave speed and friction, with the following considerations:

Field measurements clearly indicated the evolution of the transient. The pressure wave speed for pipeline with typical material and bedding was determined based on the period of the transient ($4\,L/a$) and the length (L) between measurement locations. There was air in the system; the measured wave speed was much lower than the theoretical speed. Pressure waves did not propagate through an established mixture of liquid and vapor bubbles; this inability distinguished vaporous cavitations from gaseous cavitations. Both the collapse of a large vapor cavity and the movement of the shock wave front into a vaporous cavitations zone made the vapor condense back to liquid. Typically, numerical schemes did not reproduce the exact timing of column separation events.

The model was heterogeneous (varying state within the system), and then the parameters are distributed. Distributed parameters were typically represented with partial differential equation. It was important to emphasize here that the purpose in applying a physically based numerical model in this research was to obtain insights into the interaction between heterogeneities in water flow structure and transient processes, rather than to make precise predictions of water hammer surge pressures [1-20]. Taking a distributed-parameter approach permits several structural attributes to be varied in a controlled fashion, including the types and distribution of material properties.

Two cases are considered for modeling [12-27]:

1. The inlet pressure of the pipe length is equal to p_0. The slugging pressure has a sharp increasing: $\Delta p_{y\partial} : p = p_0 + \Delta p_{y\partial}$. The N. E. Zhukovsky formula is as flowing:

$$\Delta p_{y\partial} = (C.\Delta v / g),\qquad(1)$$

where g – acceleration of free fall. The speed of the shock wave is calculated by the formula:

$$C = \sqrt{\dfrac{g \cdot \dfrac{E_*}{\rho}}{1 + \dfrac{d}{\delta} \cdot \dfrac{E_*}{E}}},\qquad(2)$$

where E_{α} – modulus of elasticity of the liquid (water) $MR = a + bt + ct^2$ $\left(\frac{kg}{m^2}\right)$
E – modulus of elasticity for pipeline material Steel $E = 10^{11}(Pa), \left(\frac{kg}{m^2}\right), d$
– outer diameter of the pipe (mm), ρ – density of the liquid (water) $\left(\frac{kg}{m^3}\right)$,
δ – pipe wall thickness (mm)
Stopping of a second layer of liquid exerts pressure on the following layers gradually caused high pressure. It acts directly at the valve extends to the rest of the pipeline against fluid flow speed C.

2. The method of characteristics MOC is defined based on a finite difference technique where pressures are computed along the pipe for each time step,

$$\left(P_1 / \gamma\right) + Z_1 + \left(V_1^2 / 2g\right) + h_p = \left(P_2 / \gamma\right) + Z_2 + \left(V_2^2 / 2g\right) + h_L,\qquad(3)$$

$$(g / a)(dH / dt) + dv / dt + \left(f\, v|v|2d\right) = 0 \Rightarrow (ds / dt) = c^+,\qquad(4)$$

$$-(g / a)(dH / dt) + dv / dt + \left(f\, v|v|2d\right) = 0 \Rightarrow (ds / dt) = c^-,\qquad(5)$$

The method of characteristics and finite difference technique compute shock wave along the pipe for each time step. Calculation automatically subdivided the pipe into sections (intervals) and selected a time interval for computations.

$$(dp / dt) = (\partial p / \partial t) + (\partial p / \partial s)(ds / dt), \tag{6}$$

$$(dv / dt) = (\partial v / \partial t) + (\partial v / \partial s)(ds / dt), \tag{7}$$

P and V changes due to time are high and due to coordination are low then it can be neglected for coordination differentiation:

$$(\partial v / \partial t) + (1 / \rho)(\partial p / \partial s) + g(dz / ds) + (f / 2D)v|v| = 0,$$

$$\text{(Euler equation), (8)}$$

$$C^2(\partial v / \partial s) + (1 / P)(\partial P / \partial t) = 0, \text{ (Continuity equation),} \tag{9}$$

By linear combination of Euler and continuity equations in characteristic solution Method:

$$\lambda\left[(\partial v / \partial t) + (1 / \rho)(\partial p / \partial s) + g(dz / ds) + (f / 2D)v|v|\right] + C^2(\partial v / \partial s) + (1 / p)(\partial p / \partial t) = 0,$$
$$\lambda = {}^+ c \,\&\, \lambda = {}^- c$$

$$\tag{10}$$

$$(dv / dt) + (1 / cp)(dp / ds) + g(dz / ds) + (f / 2D)v|v| = 0, \tag{11}$$

$$(dv / dt) - (1 / cp)(\partial p / \partial s) + g(dz / ds) + (f / 2D)v|v| = 0, \tag{12}$$

Method of characteristics drawing in (s-t) coordination:

$$(dv / dt) - (g / c)(dH / dt) = 0, \tag{13}$$

$$dH = (c \, / \, g) dv, \text{ (Joukowski Formula)}, \tag{14}$$

By Finite Difference method:

$$c+ : \left((vp - v_{Le})(Tp - 0) \right) + \left((g \, / \, c)(Hp - H_{Le}) \, / \, (Tp - 0) \right) + \left(\left(fv_{Le} \middle| \, v_{Le} \middle| \right) / \, 2D \right) = 0 \middle|$$

$$\tag{15}$$

$$c- : \left((vp - vRi)(Tp - 0) \right) + \left((g \, / \, c)(Hp - HRi) \, / \, (Tp - 0) \right) + \left(\left(fvRi \middle| v \, Ri \middle| \right) / \, 2D \right) = 0 ,$$

$$\tag{16}$$

$$c+ : (vp - v_{Le}) + (g \, / \, c)(Hp - H_{Le}) + (f \Delta t) \left(fv_{Le} \middle| v_{Le} \middle| \right) / \, 2D = 0 \quad , \tag{17}$$

$$c- : (vp - vRi) + (g \, / \, c)(Hp - HRi) + (f \Delta t) \left(v_{Ri} \middle| v_{Ri} \middle| \right) / \, 2D = 0 , \tag{18}$$

$$V_P = 1/2 \begin{pmatrix} (V_{Le} + V_{ri}) + (g \, / \, c)(H_{Le} - H_{ri}) \\ -(f \, \Delta t \, / \, 2D)(V_{Le} \, \middle| V_{Le} \middle| + V_{ri} \middle| V_{ri} \middle|) \end{pmatrix}, \tag{19}$$

$$H_P = 1/2 \begin{pmatrix} C \, / \, g (V_{Le} - V_{ri}) + (H_{Le} + H_{ri}) \\ -C \, / \, g (f \, \Delta t \, / \, 2D)(V_{Le} \, \middle| V_{Le} \middle| - V_{ri} \middle| V_{ri} \middle|) \end{pmatrix}, \tag{20}$$

A model for liquid-vapor flows illustrates the numerical techniques for solving the resulting equations. Hence field test model on actual systems was chosen for experimental presentation of water hammer phenomenon. This work although included the description of air entrance phenomenon at the flow discontinuities and changes for gas content [28-42].

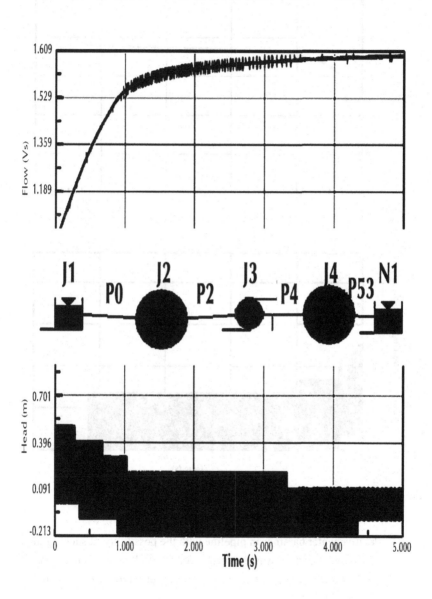

FIGURE 1   Laboratory model for air entrance for water hammer phenomenon.

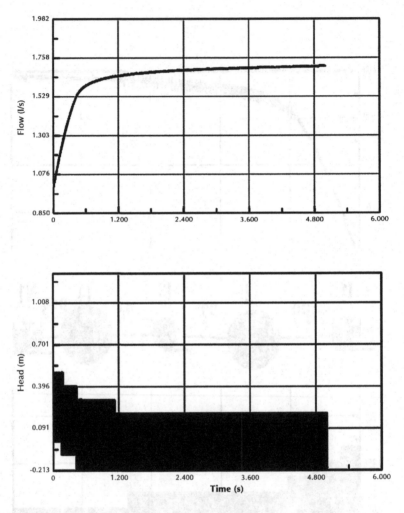

FIGURE 2 Laboratory model for liquid-vapor flows.

- Positive water hammer in pipeline with the possibility of sucking in air in negativephase. This was the reason for the sucking air in negative phase for Water Pipeline. Consistence between observed values of maximal pressure in first.
- Significant influence of the rate of the discharge decreased the duration of the water hammer phenomenon. The duration time decreased with the

increase of the outflow from the discharge. This was the strong reason for the high deceasing in duration time for water pipeline [43–67].

7.2.1   NEWTON SECOND LAW FOR RESEARCH LABORATORY MODEL(21–27)

$$\rho a l \frac{dv}{dt} = \rho g a H_1 - \rho g a (H_2 + y) + \rho g a L \sin\theta - \rho g a h_f , \qquad (21)$$

$$\frac{L}{g} \times \frac{dv}{dt} + y + h_f = 0 \ a.v = A\frac{dy}{dy} + Q \ H_2 = H_1 + k \ L\sin\theta = k , \qquad (22)$$

$$\frac{L}{g} \times \frac{d}{dt}\left(\frac{A}{a}\frac{dy}{dt} + \frac{Q}{a} \right) + y + h_f = 0 , \qquad (23)$$

$$\frac{d^2y}{dt^2} + \frac{ga}{LA} y = 0 , \qquad (24)$$

$$\frac{d^2y}{dt^2} + W^2 y = 0 , \qquad (25)$$

$$\frac{L}{g} \times \frac{d}{dt}\left(\frac{A}{a}\frac{dy}{dt} + \frac{Q}{a} \right) + y + h_f = 0 , \qquad (26)$$

$$\frac{d^2y}{dt^2} + \frac{ga}{LA} y = 0 , \qquad (27)$$

C$_v$, Discharge Coefficient (m2.5/s), Area (m^2), Degree (rad), Diameter (m), Elevation or Head (m), Flow (l/s), Force (N), Inertia (N-m^2), Length (m), Pipe Diameter, Rou (mm), Pressure (m-Hd), Rotational Speed (rpm), Roughness Factor C (HW), Specific Speed (rpm 0.75 m/0.5 s), Spring Constant (N/mm), Time(s), Torque (Nm), Velocity (m/s), Viscosity (m^2/s), Volume (m^3).

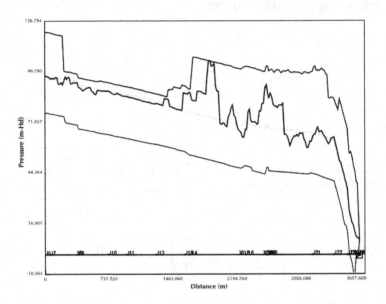

FIGURE 3   Simulation for Maxpressure variation.

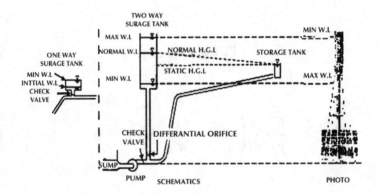

FIGURE 4   Laboratory model for water hammer.

7.2.2 SECOND MODEL APPROACHES TO TRANSIENT FLOW (METHOD OF CHARACTERISTICS "MOC" MODEL)

Second Model was formed by Method of characteristics "MOC." So Specification and Geography (geo reference coordinators) of System were defined based on Geography Information System "GIS." Specification and Geography of System as an input data file were interred to Water Hammer Software, Version 07.00.049.00. Second Model results for reducing on pressure drop related to water transmission rate were compared by first Model. The outputs were included output data files and output curves. Transient flow curves were included: Min Head-Distance; Elevation-Distance; Max Volume-Distance; Pressure-Distance; Max Pressure-Distance curves; Flow-Time; Volume-Time; Head-Time; Pressure–Time Transient curves results achieved in Refs. [68-77]. Present work by comparison of Laboratory Model and "MOC" Model found the effects of pumps numbers and Two-Way Surge Tank on pressure drop at water hammer condition. Thus work found optimum working point of pumps at water hammer condition for water Transmission Line system.

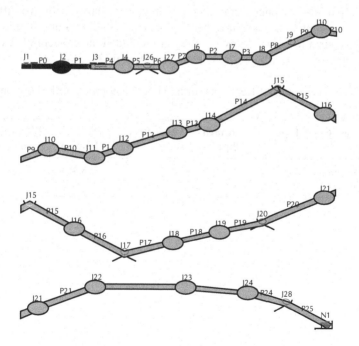

FIGURE 5   Simulation of pipeline; field test model.

In This work a two-way surge tank controlled transients by converting stored potential energy in the elevated water body inside the tank into kinetic energy, which supplements flow in the piping system at critical times (or vice versa, for pipe flow into the tank) during periods of rapid flow variation. The tank was located at the pumping station.

A differential orifice was installed at the riser of the tank to throttle reverse flow from the system to the tank, but created very little loss for flow leaving the tank. If an overflow and drain was provided, the tank could also act as a foolproof over pressure device that could overflow in a controlled manner. One of the main concerns was the stability problem inside the tank. A rapid rise or drop in water level in the tank should be avoided. Usually, the surface area of the tank should be significantly larger than that of the pipeline. In a high-head water system or a sanitary force main, a two-way surge tank may not be economically feasible because of height or odor problems [78-97].

7.3   RESULTS AND DISCUSSION

Modeling of air influence (Table 1) on hydraulic similarity with two different types of air content models have been proposed in the literature in predicting of the transient pressure behavior. They are included the concentrated vaporous cavity model and the discrete air release model [33, 36].

TABLE 1   The results of field tests for maximum amount of air infiltration.

Category	Type	Branch Pipes	Vapor Pressure	Max. Volume	Type of Volume
Protection equip	Airvalve	2	-10	198.483	Air

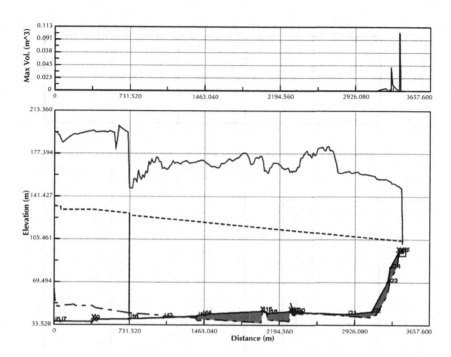

FIGURE 6 Water column separation and entered air simulation.

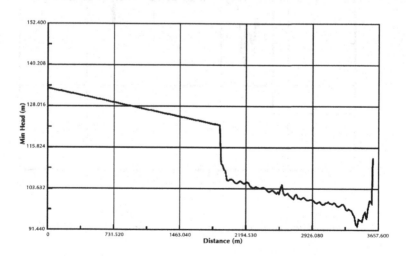

FIGURE 7 Simulation of pipeline for pressure drop.

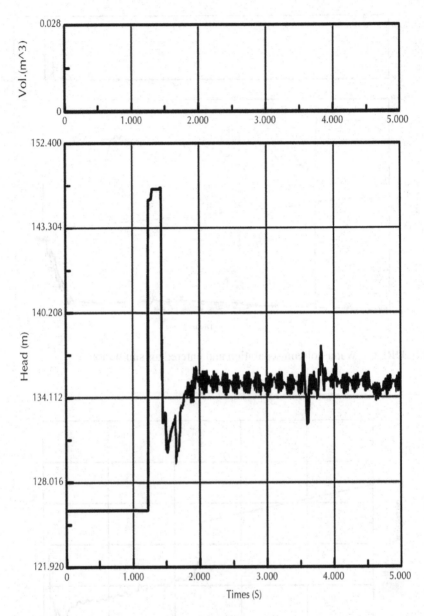

FIGURE 8　Simulation of pipeline for pressure drop changes due to flow rates variation.

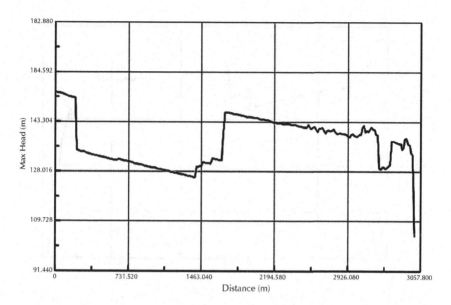

FIGURE 9   Simulation of pipeline for pressure drop changes; field test model.

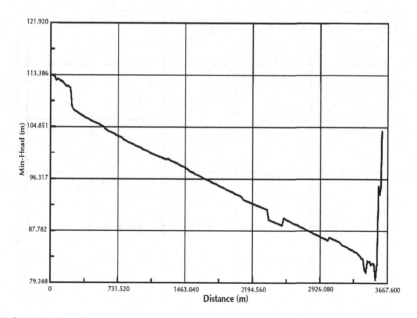

FIGURE 10   Simulation of pipeline.

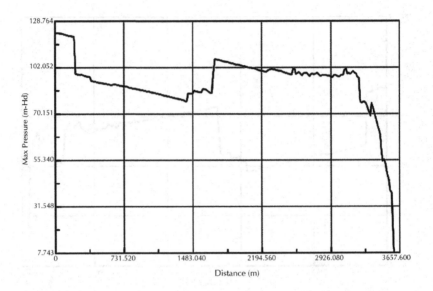

FIGURE 11   Simulation for Maxpressure variation.

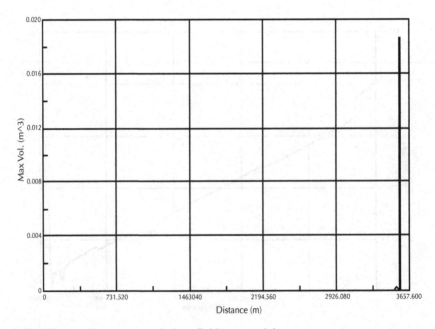

FIGURE 12   Flow rates variation, field test model

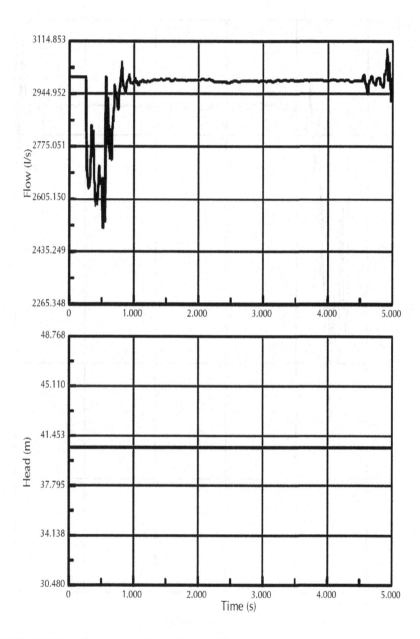

FIGURE 13   Simulation of pipeline for flow rates variation

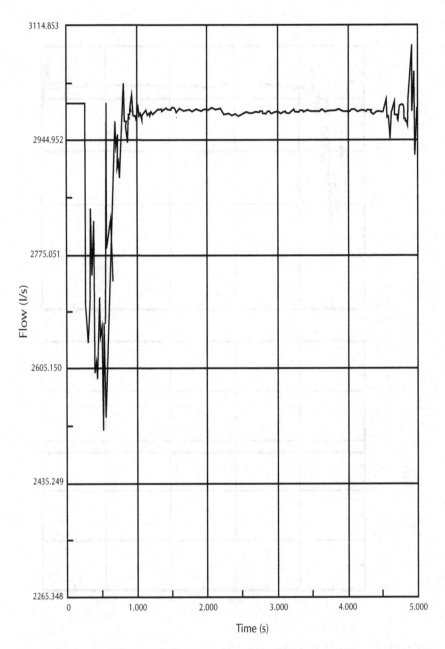

FIGURE 14   Simulation of flow rates variation, field test model.

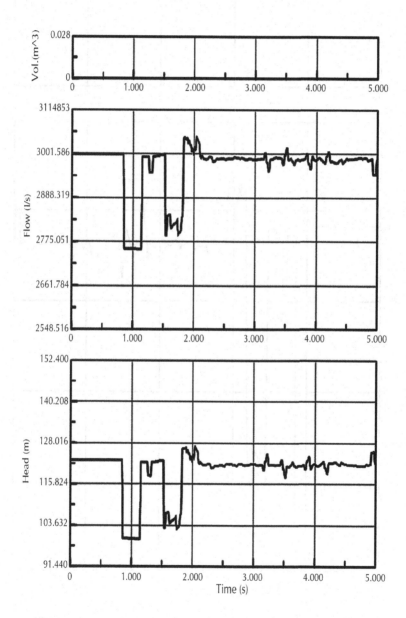

FIGURE 15   Pressure and flow changes in pipeline.

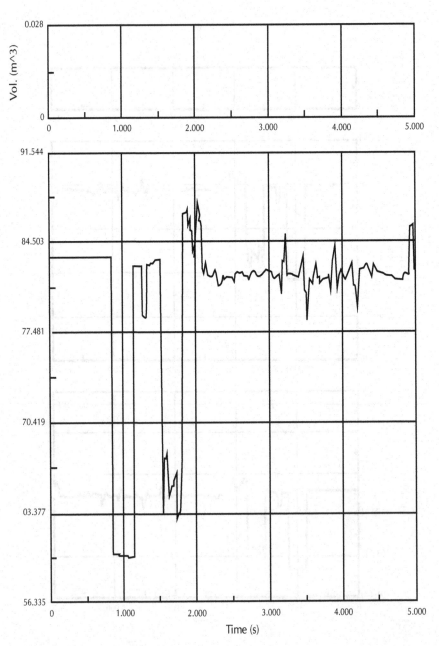

FIGURE 16   Field test Model for pressure changes.

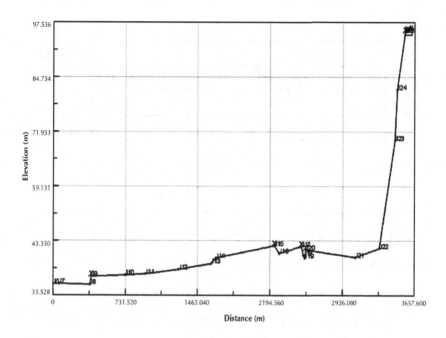

FIGURE 17 Elevation changes of water pipeline.

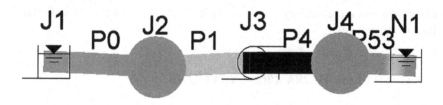

FIGURE 18 Simulation of laboratory model.

7.3.1 AIR ENTRANCE APPROACHES

Analysis of the nonlinear heterogeneous model showed that air penetrated (Fig.19) into water pipeline (Table 1).

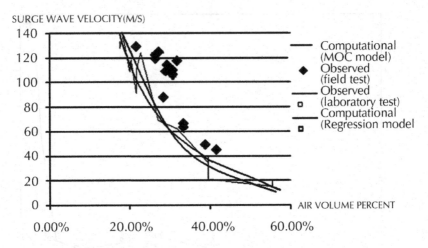

FIGURE 19   Laboratory experiments and field-test results.

7.3.2 COMPARISON OF PRESENT WORK RESULTS WITH OTHER EXPERT'S RESEARCH

Comparison of present work results (Figs. 20–33) with the results of other expert's works on laboratory experiments and field test results shows similarity and advantages.

7.3.3 SIMILAR WORK PRESENTATIONS

7.3.3.1 PRESENT RESEARCH MODELING CAPABILITIES

The differences between computer model results and actual system measurements were caused by several factors, including the following difficulties:

1. Precise determination of the pressure wave speed for the piping system was difficult, if not impossible. This was especially true for buried pipelines, whose wave speeds were influenced by bedding conditions and the compaction of the surrounding soil.

2. Precise modeling of dynamic system elements (such as valves, pumps, and protection devices) was difficult because they were subject to deterioration with age and adjustments made during maintenance activities. Measurement equipment may also be inaccurate.

3. Unsteady or transient friction coefficients and losses depend on fluid velocities and accelerations. These were difficult to predict and calibrate even in laboratory conditions.

4. Prediction of the presence of free gases in the system liquid was sometimes impossible.

These gases can significantly affect the pressure wave speed. In addition, the exact timing of vapor-pocket formation and column separation was difficult to simulate. Calibrating model parameters based on field data minimized the first source of error.

Unsteady or transient friction coefficients and the effects of free gases are more challenging to account for. Fortunately, friction effects are usually minor in most water systems and vaporization can be avoided by specifying protection devices and/or stronger pipes and fittings able to withstand sub atmospheric or vacuum conditions, which are usually short-lived.

For present research with free gas systems and the potential for water-column separation, the numerical simulation of hydraulic transients was more complex. Small pressure spikes caused by the type of tiny vapor pockets that was difficult to simulate accurately seldom result in a significant change to the transient envelopes [98-107]. Larger vapor-pocket collapse events resulting in significant upsurge pressures were simulated with enough accuracy to support definitive conclusions.

With respect to timing, there were close agreement between the Computed and measured periods of the system, regardless of what flow-control operation initiated the transient. With a well-calibrated model of the system, it was possible to use the model in the operational control of the system and anticipated the effects of specific flow-control operations. This required field measurements to quantify system's pressure wave speed and friction, with the following considerations:

• Field measurements clearly indicated the evolution of the transient. The pressure wave speed for pipeline with typical material and bedding was determined based on the period of the transient ($4\ L/a$) and the length (L) between measurement locations. There was air in the system; the measured wave speed was much lower than the theoretical speed. Pressure waves did not propagate through an established mixture of liquid and vapor bubbles; this inability distinguished vaporous cavitations from gaseous cavitations. Both the collapse of a large vapor cavity and the movement of the shock wave front into a vaporous cavitations zone made the vapor condense back to liquid. Typically, numerical schemes did not reproduce the exact timing of column separation events.

The model was heterogeneous (varying state within the system), and then the parameters are distributed. Distributed parameters were typically represented with partial differential equation. It was important to emphasize here that the purpose in applying a physically based numerical model in this research was to obtain insights into the interaction between heterogeneities in water flow structure and transient processes, rather than to make precise predictions of water hammer surge pressures. Taking a distributed-parameter approach permits several structural attributes to be varied in a controlled fashion, including the types and distribution of material properties. This systems was somewhere between the black-box (a system of which there is no a priori information available) and white-box models (a system where all necessary information is available). Transient vaporous cavitations (including column Separation) occurred in pipelines when the liquid pressure falls to the liquid vapor pressure. The model used in this research was MOC. The model was solved the integral-differential equation.

$$Pressure = 28.762 + .031 Flow - .005 Dis\tan ce + .731 Time \quad (28)$$

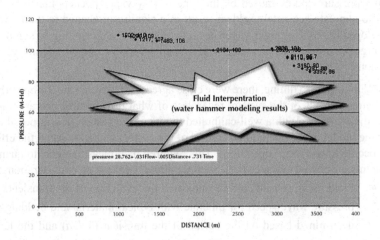

FIGURE 20   Fluid interpenetration modeling for water pipeline.

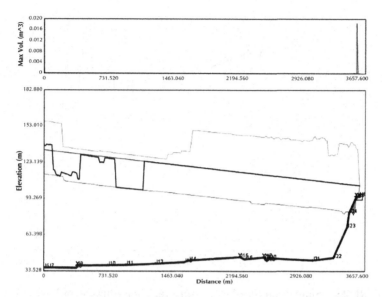

FIGURE 21   Simulation for Max pressure variation (first record).

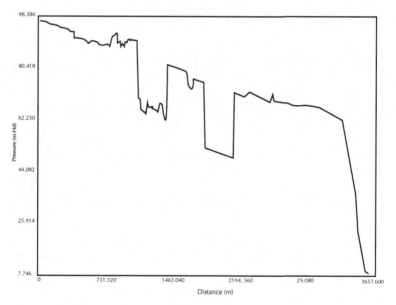

FIGURE 22   Simulation for Max pressure variation (second record).

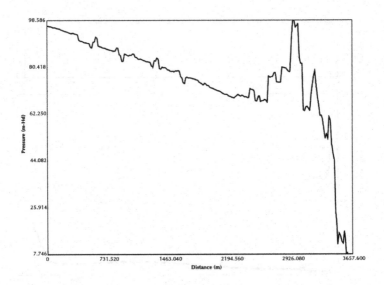

FIGURE 23   Simulation for Max pressure variation (third record).

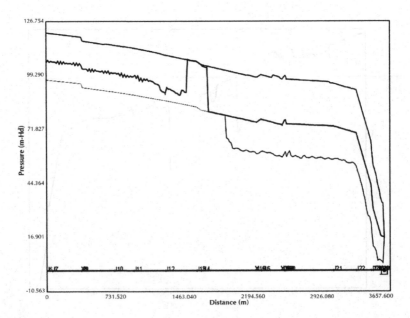

FIGURE 24   Simulation for Max pressure variation (forth record).

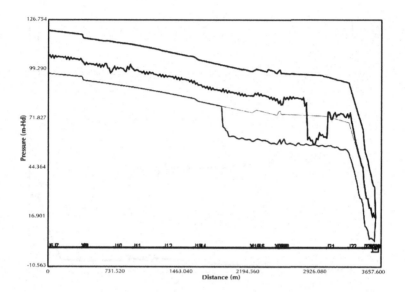

FIGURE 25   Simulation for Max pressure variation (fifth record).

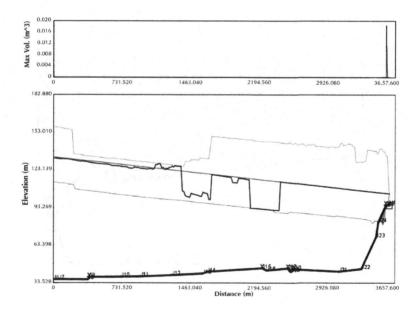

FIGURE 26   Simulation for Max pressure variation (sixth record).

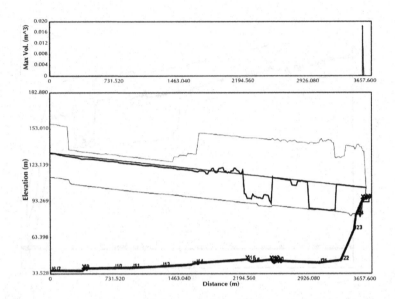

FIGURE 27   Simulation for Max pressure variation (seventh record).

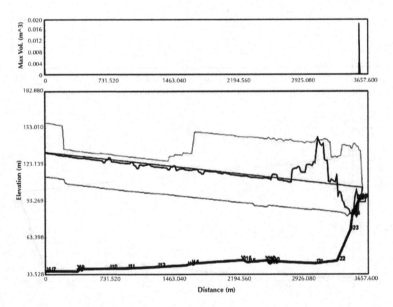

FIGURE 28   Simulation for Max pressure variation (eighth record).

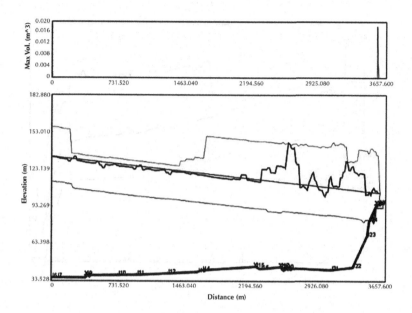

FIGURE 29   Simulation for Max pressure variation (ninth record).

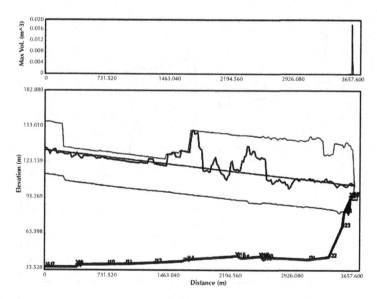

FIGURE 30   Simulation for Max pressure variation (tenth record).

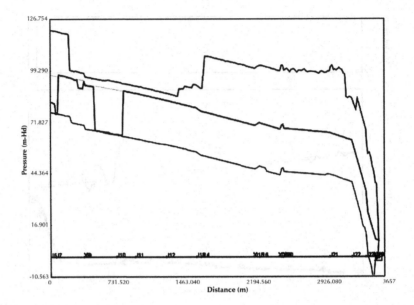

FIGURE 31   Simulation for Max pressure variation (eleventh record).

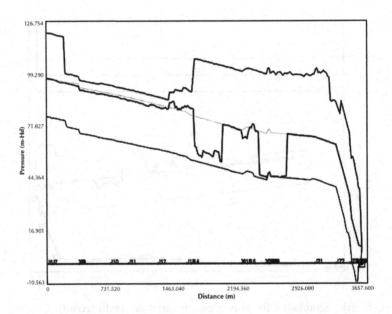

FIGURE 32   Simulation for Max pressure variation (twelfth record).

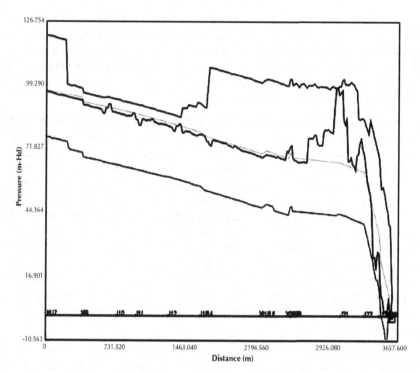

FIGURE 33   Simulation for Max pressure variation (thirteenth record).

7.3.4   LOCAL LEAKAGE RATE EFFECT ON THE MAXIMUM VALUE OF PRESSURE

The reason for the high-pressure drop in water transmission was in relation with the discharge in water pipeline). This work led to improved standards for precession designs and installation techniques in the field of sub atmospheric transient pressures that can suck contaminants into the water system.

A treated or modeled air entrainment problem in real prototype systems and results was showed in. Consistency between the observed values of maximal pressure in the first amplitude and corresponding values were calculated according to Joukowski's formula, irrespective of the rate of flow. Significant influence of the discharge rate into the pipeline decreases of duration of the water hammer phenomenon.

The duration time decreased with the increase of the air penetration. This was the strong reason for the high deceasing in duration time for water pipeline. In this work the column separations due to the turned off pump for the water pipeline and the air penetration were carried out. Pressure at the beginning of pipeline remains constant. After shock wave generation, flow moves with the same velocity C at the reverse direction of the shock wave.

This leads to the generation of high-pressure drop due to the wave. At the same time fluid moves in the reverse direction of the pipeline. As long as the shock wave reaches to the pressure reducing valve, liquid pressure reduces to vapor pressure. By the way, again and again, the wave of pressure drop moves conversely in the direction of start point on the water pipeline. As long as damping of shock wave, these cycles of increase and decrease of pressure will be continued. It is iterated at time intervals equal to time for dual-path of the shock wave along with the length of the pipeline (from the pressure reducing valve prior to the start point of pipeline). The hydraulic impact of the liquid in the pipeline will perform oscillatory motion. The hydraulic resistance and viscosity cause the oscillatory motion. It absorbs the initial energy of the liquid as long as overcoming the friction and therefore it will be damped. Water hammer is manifested in hydro-machines various purposes. In most cases this is undesirable, leading to the destruction of pipelines.

Maximum amount of air infiltration, which was calculated, based on the simulation results of nonlinear heterogeneous model were released by the air relief valve on the system.

7.4 CONCLUSIONS

Detailed conclusions were drawn on the basis of experiments by Laboratory model and calculations for the pipeline with air entered. Hence the most important effects that have been observed were as the flowing: at first, the influences of the ratio of air have been appointed. Then the total discharge related to the periodic wave oscillations has been investigated. The pipeline was equipped with the valve at the end of the main pipe, which was joined with the closure time register.

KEYWORDS

- **Air relief valve**
- **Computational method**
- **Shock wave**
- **Surge tank**
- **Water column separation**

REFERENCES

1. Hariri Asli, K.; GIS Water hammer disaster at earthquake in Rasht water pipeline, 3rd International Conference on Integrated Natural Disaster Management, INDM **2008,** http://www.civilica.com/Paper-INDM03-INDM03_001.html.

2. Hariri Asli, K.; Nagiyev, F. B.; Beglou, M. J.; Haghi, A. K.; Kinetic analysis of convective drying, International Journal of the Balkan Tribological Association, ISSN: 1310–4772, Sofia, Bulgaria, **2009,** *15, 4,* 546–556, jbalkta@gmail.com

3. Hariri Asli, K.; Nagiyev, F. B.; Bubbles characteristics and convective effects in the binary mixtures. Transactions issue mathematics and mechanics series of physical-technical and mathematics science, ISSN 0002–3108, Azerbaijan, Baku, **2008,** 215–220, www.imm.science.az/journals/AMEA_xeberleri/.../215–220.pdf

4. Hariri Asli, K.; Nagiyev, F. B.; Haghi, A. K.; Three-dimensional Conjugate Heat Transfer in Porous Media, International Journal of the Balkan Tribological Association, ISSN: 1310–4772, Sofia, Bulgaria, *15, 3,* 336–346, **2009,** jbalkta@gmail.com

5. Hariri Asli, K.; Nagiyev, F. B.; Haghi, A. K.; Water hammer and fluid condition; a computational approach, Computational Methods in Applied Science and Engineering, USA, Chapter *5,* Nova Science Publications, ISBN: 978-1-60876-052-7, USA, 73–94, **2010,** https://www.novapublishers.com/catalog/

6. Hariri Asli, K.; Nagiyev, F. B.; Haghi, A. K.; Interpenetration of two fluids at parallel between plates and turbulent moving in pipe; a case study, Computational Methods in Applied Science and Engineering, USA, Chapter *7,* Nova Science Publications, ISBN: 978-1-60876-052-7, USA, 107–133, 2010, https://www.novapublishers.com/catalog/

7. Hariri Asli, K.; Nagiyev, F. B.; Haghi, A. K.; Modeling for water hammer due to valves; from theory to practice, Computational Methods in Applied Science and Engineering, USA, Chapter 11, Nova Science Publications ISBN: 978-1-60876-052-7, USA, 229–236, **2010,** https://www.novapublishers.com/catalog/

8. Hariri Asli, K.; Nagiyev, F. B.; Haghi, A. K.; A computational method to Study transient flow in binary mixtures, Computational Methods in Applied Science and En-

gineering, USA, Chapter 13, Nova Science Publications ISBN: 978-1-60876-052-7, USA, 229–236, **2010**, https://www.novapublishers.com/catalog/

9. Hariri Asli, K.; Nagiyev, F. B.; Haghi, A. K.; Water hammer analysis; some computational aspects and practical hints, Computational Methods in Applied Science and Engineering, USA, Chapter 16, Nova Science Publications ISBN: 978-1-60876-052-7, USA, 263–281, **2010**, https://www.novapublishers.com/catalog/

10. Hariri Asli, K.; Nagiyev, F. B.; Haghi, A. K.; Water hammer and hydrodynamics instabilities modeling, Computational Methods in Applied Science and Engineering, USA, Chapter 17, From Theory to Practice, Nova Science Publications ISBN: 978-1-60876-052-7, USA, 283–301, **2010**, https://www.novapublishers.com/catalog/

11. Hariri Asli, K.; Nagiyev, F. B.; Haghi, A. K.; A computational approach to study water hammer and pump pulsation phenomena, Computational Methods in Applied Science and Engineering, USA, Chapter 22, Nova Science Publications, ISBN: 978-1-60876-052-7, USA, 349–363, **2010**, https://www.novapublishers.com/catalog/

12. Hariri Asli, K.; Nagiyev, F. B.; Haghi, A. K.; Some aspects of physical and numerical modeling of water hammer in pipelines. Computational Methods in Applied Science and Engineering, USA, Chapter 23, Nova Science Publications, ISBN: 978-1-60876-052-7, USA, 365–387, **2010**, https://www.novapublishers.com/catalog/

13. Hariri Asli, K.; Nagiyev, F. B.; Haghi, A. K.; A computational approach to study fluid movement, Nanomaterials Yearbook – **2009**, From Nanostructures, Nanomaterials and Nanotechnologies to Nanoindustry, Chapter 16, Nova Science Publications, USA, ISBN: 978-1-60876-451-8, USA, 181–196, 2010, https://www.novapublishers.com/catalog/product_info.php?products_id=11587

14. Hariri Asli, K.; Nagiyev, F. B.; Haghi, A. K.; Physical modeling of fluid movement in pipelines, Nanomaterials Yearbook – **2009**, From Nanostructures, Nanomaterials and Nanotechnologies to Nanoindustry, Chapter 17, Nova Science Publications, USA, ISBN: 978-1-60876-451-8, USA, 197–214, 2010, https://www.novapublishers.com/catalog/product_info.php?products_id=11587

15. Hariri Asli, K.; Nagiyev, F. B.; Haghi, A. K.; Aliyev, S. A.; Improved modeling for prediction of water transmission failure, Recent Progress in Research in Chemistry and Chemical Engineering, Chapter 2, Nova Science Publications, ISBN: 978-1-61668-501-0, Nova Science Publications, USA, 28–36, **2010**, https://www.novapublishers.com/catalog/product_info.php?products_id=13174

16. Hariri Asli, K.; Nagiyev, F. B.; Haghi, A. K.; Aliyev S. A.; Pure Oxygen penetration in wastewater flow, Recent Progress in Research in Chemistry and Chemical Engineering, Chapter 3, Nova Science Publications, ISBN: 978-1-61668-501-0, Nova Science Publications, USA, 17–27, **2010**, https://www.novapublishers.com/catalog/product_info.php?products_id=13174

17. Hariri Asli, K.; Mathematics and numerical modeling Technology, Journal of Mathematics and Technology, ISSN: 2078–0257, No.3, August, Baku, Azerbaijan, 68–74, **2010**,   https://www.International%20Journal%20of%20Academic%20Research-IJAR. htm

18. Hariri Asli, K.; Nagiyev, F. B.; Haghi, A. K.; Aliyev, S. A.; Physical and Numerical Modeling of Fluid Flow in Pipelines: A computational approach, International Journal of the Balkan Tribological Association, ISSN: 1310–4772, 16, *1*, Sofia, Bulgaria, 20–34, **2010**, jbalkta@gmail.com

19. Hariri Asli, K.; Nagiyev, F. B.; Haghi, A. K.; Aliyev, S. A.; A Numerical Study on heat transfer in Microtubes, International Journal of the Balkan Tribological Association, ISSN: 1310–4772, 16, *1,* Sofia, Bulgaria, **2010,** 9–19, jbalkta@gmail.com

20. Hariri Asli, K.; Nagiyev, F. B.; Haghi, A. K.; A numerical study on fluid dynamics, Material Science Synthesis, Properties, Applicators, ISBN: 978-1-60876-872-1, Chapter 15, Nova Science Publications, USA, **2010,** 101–110. https://www.novapublishers.com/catalog/product_info.php?products_id=12129

21. Hariri Asli, K.; Nagiyev, F. B.; Haghi, A. K.; Some interpenetration for turbulent moving of fluid in pipe, Material Science Synthesis, Properties, Applicators, ISBN: 978-1-60876-872-1, Chapter 16, Nova Science Publications, USA, 111–117, **2010,** https://www.novapublishers.com/catalog/product_info.php?products_id=12129

22. Hariri Asli, K.; Nagiyev, F. B.; Haghi, A. K.; Fluid flow analysis due to water hammer, Material Science Synthesis, Properties, Applicators, ISBN: 978-1-60876-872-1, Chapter 17, Nova Science Publications, USA, **2010,** 120–128. https://www.novapublishers.com/catalog/product_info.php?products_id=12129

23. Hariri Asli, K.; Nagiyev, F. B.; Haghi, A. K.; Transient flow in binary mixtures, Material Science Synthesis, Properties, Applicators, ISBN: 978-1-60876-872-1, Chapter 19, Nova Science Publications, USA, **2010,** 164–176. https://www.novapublishers.com/catalog/product_info.php?products_id=12129

24. Hariri Asli, K.; Nagiyev, F. B.; Haghi, A. K.; Hydrodynamics instabilities modeling, Material Science Synthesis, Properties, Applicators, ISBN: 978-1-60876-872-1, Chapter 20 Nova Science Publications, USA, **2010,** 140–146. https://www.novapublishers.com/catalog/product_info.php?products_id=12129

25. Hariri Asli, K.; Nagiyev, F. B.; Haghi, A. K.; Fluid dynamics and pump pulsation, Material Science Synthesis, Properties, Applicators, ISBN: 978-1-60876-872-1, Chapter 21, Nova Science Publications, USA, **2010,** 147–155. https://www.novapublishers.com/catalog/product_info.php?products_id=12129

26. Hariri Asli, K.; Nagiyev, F. B.; Haghi, A. K.; Aliyev, S. A.; Hariri Asli, H.;Flow in water pipeline: A computational approach, International Journal of Academic Research, ISSN: 1310–4772, ISSN: 2075–4124, *2,* Issue *5,* September30, Baku, Azerbaijan, **2010,** 164–176, https://www.International%20Journal%20of%20Academic%20Research-IJAR. htm

27. Hariri Asli, K.; Nagiyev, F. B.; Haghi, A. K.; Aliyev, S. A.; Nonlinear Heterogeneous Model for Water Hammer Disaster, International Journal of the Balkan Tribological Association, ISSN: 1310–4772, *16, 2,* Sofia, Bulgaria, 209–222, **2010,** jbalkta@gmail.com

28. Hariri Asli, K.; Nagiyev, F. B.; Haghi, A. K.; Heat flow and mass transfer in capillary Porous body, Journal of the Balkan Tribological Association, *16(3),* Tribotechnics and tribomechanics, Sofia, Bulgaria, 353–361, **2010,** jbalkta@gmail.com

29. Hariri Asli, K.; Nagiyev, F. B.; Haghi, A. K.; A Numerical Study on thermal drying of Porous solid, Journal of the Balkan Tribological Association, *16(3),* Tribotechnics – thermal drying, Sofia, Bulgaria, 373–381, **2010,** jbalkta@gmail.com

30. Hariri Asli, K.; Haghi, A. K.; A Numerical Study on Fluid Flow and Pressure drop in Microtubes, Journal of the Balkan Tribological Association, *16(3),* Tribotechnics and tribomechanics, Sofia, Bulgaria, 382–392, **2010,** jbalkta@gmail.com

31. Hariri Asli, K.; Nagiyev, F. B.; Haghi, A. K.; Aliyev, S. A.; Hariri Asli, H.;Improved Nonlinear Heterogeneous Model for Wastewater Treatment, International Journal on "Technical and Physical Problems of Engineering," (IJTPE), Published by the International Organization on TPE (IOTPE), ISSN: 2077–3528, Baku, Azerbaijan, 30–36, **2010,** http://www.iotpe.com/TPE-Journal/PublicationPolicy.html

32. Hariri Asli, K.; GIS Nonlinear Dynamics Model: Some Computational Aspects and Practical Hints, International Journal on "Technical and Physical Problems of Engineering," (IJTPE), Published by the International Organization on TPE (IOTPE), ISSN: 2077–3528, 1–5, Baku, Azerbaijan, **2010,** http://www.iotpe.com/TPE-Journal/PublicationPolicy.html

33. Hariri Asli, K.; Numerical Modeling for Transient Flow; Some Engineering Aspects and Economical Hints, Journal of Economics and Engineering, ISSN: 2078–0346, *3,* August, 18–24, Baku, Azerbaijan, **2010,** https://www.International%20Journal%20of%20Academic%20Research-IJAR. htm

34. Hariri Asli, K.; Nagiyev, F. B.; Haghi, A. K.; Aliyev, S. A.; Numerical Modeling for water hammer in pipeline; from theory to practice, Journal of Mechanics Machine Building, ISSN: 1816–4986, *1,* August, 43–52, Baku, Azerbaijan, 2010.

35. Hariri Asli, K.; Nagiyev, F. B.; Haghi, A. K.; "Some Aspects of Physical and Numerical Modeling of Water Hammer in Pipelines," Nonlinear DynamicsAn International Journal of Nonlinear Dynamics and Chaos in Engineering Systems, ISSN: 1573–269X (electronic version) Journal no. 11071 Springer, Netherlands, **2009,** ISSN: 0924–090X (print version), Springer, Heidelberg, Germany, Volume *60,* Number 4 / June, 677–701, **2010,** http://www.springerlink.com/openurl.asp?genre=article&id=doi: 10.1007/s11071–009–9624–7.

36. Haghi, A. K.; Hariri Asli, K.; Application of nano-SiO2 in cement composites, Journal of the Balkan Tribological Association, *16, 4,* Tribotechnics and tribomechanics, Sofia, Bulgaria, 585–594, **2010,** jbalkta@gmail.com

37. Haghi, A. K.; Hariri Asli, K.; E.; Sabermaash, A Review on Electrospun Polymeric Nanosized Fibers, Journal of the Balkan Tribological Association, *16, 4,* Tribotechnics and tribomechanics, Sofia, Bulgaria, 570–584, **2010,** jbalkta@gmail.com

38. Hariri Asli, K.; Nagiyev, F. B.; Haghi, A. K.; Aliyev, S. A.; Hariri Asli, H.;Modeling of fluid interaction produced by water hammer, International Journal of Chemoinformatics and Chemical Engineering, 1(1), ISSN: 2155–4110, eISSN: 2155–4129, USA, **2011,** 29–41, http://64.225.152.8/proofs/IJCCE/IJCCE. pdf

39. Hariri Asli, K.; Haghi, A. K.; Potentials for Use of Recycled Rubber-Tire Particles as Concrete Aggregate, Journal of the Balkan Tribological Association, *17, 2,* Tribotechnics – thermal drying, Sofia, Bulgaria, 319–326, 2011, jbalkta@gmail.com

40. Hariri Asli, K.; Haghi, A. K.; Hariri Asli, H.;Sabermaash Eshghi E.; Improved Modeling for Water Surge Wave, The International Journal of Multimedia Technology, 7–13, Pub. Date: 2011–09–30, ISSN: 2225–1456, *1,* No.1, China, **2011,** www.ijmt. org/Issue.aspx?Vol=1&Num=1&Abstr=false

41. Hariri Asli, K.; Nagiyev, F. B.; Khodaparast Haghi, R.; Hariri Asli, H.;A numerical exploration of transient decay mechanisms in water distribution systems, Advances in Control and Automation of Water Systems, Published by the Apple Academic Press, Inc.; ISSN: 978-1-926895-22-2, Toronto, Canada, **2012,** www.AppleAcademicPress. com

42. Hariri Asli, K.; Nagiyev, F. B.; Khodaparast Haghi, R.; Hariri Asli, H.;Mathematical modeling of hydraulic transients in simple systems, Advances in Control and Automation of Water Systems, Published by the Apple Academic Press, Inc.; ISSN: 978-1-926895-22-2, Toronto, Canada, **2012,** www.AppleAcademicPress.com

43. Hariri Asli, K.; Nagiyev, F. B.; Khodaparast Haghi, R.; Hariri Asli, H.;Modeling one and two-phase water hammer flows, Advances in Control and Automation of Water Systems, Published by the Apple Academic Press, Inc.; ISSN: 978-1-926895-22-2, Toronto, Canada, **2012,** www.AppleAcademicPress.com

44. Hariri Asli, K.; Nagiyev, F. B.; Khodaparast Haghi, R.; Hariri Asli, H.;Water hammer and hydrodynamics' instability, Advances in Control and Automation of Water Systems, Published by the Apple Academic Press, Inc.; ISSN: 978-1-926895-22-2, Toronto, Canada, **2012,** www.AppleAcademicPress.com

45. Hariri Asli, K.; Nagiyev, F. B.; Khodaparast Haghi, R.; Hariri Asli, H.;Hadraulic Flow Control in Binary Mixtures, Advances in Control and Automation of Water Systems, Published by the Apple Academic Press, Inc.; ISSN: 978-1-926895-22-2, Toronto, Canada, **2012,** www.AppleAcademicPress.com

46. Hariri Asli, K.; Nagiyev, F. B.; Khodaparast Haghi, R.; Hariri Asli, H.;An efficient accurate shock-capturing scheme for modeling water hammer flows, Advances in Control and Automation of Water Systems, Published by the Apple Academic Press, Inc.; ISSN: 978-1-926895-22-2, Toronto, Canada, **2012,** www.Apple AcademicPress. com

47. Hariri Asli, K.; Nagiyev, F. B.; Khodaparast Haghi, R.; Hariri Asli, H.; Applied Hydraulic Transients: Automation and Advanced control, Advances in Control and Automation of Water Systems, Published by the Apple Academic Press, Inc.; ISSN: 978-1-926895-22-2, Toronto, Canada, **2012,** www.AppleAcademicPress.com

48. Hariri Asli, K.; Nagiyev, F. B.; Khodaparast HaghiR.; Hariri Asli, H.;Improved Numerical Modeling for Perturbations in Homogeneous and Stratified Flows, Advances in Control and Automation of Water Systems, Published by the Apple Academic Press, Inc.; ISSN: 978-1-926895-22-2, Toronto, Canada, **2012,** www.AppleAcademicPress. com

49. Hariri Asli, K.; Nagiyev, F. B.; Khodaparast HaghiR.; Hariri Asli, H.;Computational Model for Water Hammer Disaster, Advances in Control and Automation of Water Systems, Published by the Apple Academic Press, Inc.; ISSN: 978-1-926895-22-2, Toronto, Canada, **2012,** www.AppleAcademicPress.com

50. Hariri Asli, K.; Nagiyev, F. B.; Khodaparast HaghiR.; Hariri Asli, H.;Heat and Mass Transfer in Binary Mixtures; A Computational Approach, Advances in Control and Automation of Water Systems, Published by the Apple Academic Press, Inc.; ISBN: 978-1-926895-22-2, Toronto, Canada, **2012,** www.AppleAcademicPress.com

51. Hariri Asli, K.; Nagiyev, F. B.; Haghi, A. K.; Aliyev, S. A.; Improved Modeling for Pressure Drop in Microtubes, Advances in Control and Automation of Water Systems, Pak. J. Sci. Ind. Res., PAGE # 36–42, *55, 1,* **2012,** http://www.pjsir.org/documnts/journals/23022012052105_Binder2abstract.pdf

52. Hariri Asli, K.; Haghi, A. K.; Hariri Asli, H.;Sabermaash Eshghi E.; Water Hammer Modelling and Simulation by GIS, Hindawi Publishing Corporation, Volume **2012,** Article ID 704163, 4 pages doi: 10.1155/2012/704163, ISSN: 16875605, 16875591, *1(3),* USA, **2012,** http://www.hindawi.com/journals/mse/aip/704163/

53. Haghi, A. K.; Hariri Asli, K.; Application of nano-SiO2 in cement composites, Journal of the Balkan Tribological Association, *18, 3,* Composite materials, Sofia, Bulgaria, 454–464, **2012**, jbalkta@gmail.com

54. Hariri Asli, K.; Fluid Interpenetration and Pulsation, Water Hammer Research; Advances in Nonlinear Dynamics Modeling, PhD.; Thesis of Kaveh Hariri Asli, Toronto, Canada, Published by Apple Academic Press, Inc.; Exclusive worldwide distribution by CRC Press, a Taylor and Francis Group, Print ISBN: 9781926895314, eBook: 978-1-46-656887-7, **2013**, www.AppleAcademicPress.com

55. Hariri Asli, K.; The Interpenetration of Fluids in the Pipe and Stratified Flow in the Channel, Water Hammer Research; Advances in Nonlinear Dynamics Modeling, PhD.; Thesis of Kaveh Hariri Asli, Toronto, Canada, Published by Apple Academic Press, Inc.; Exclusive worldwide distribution by CRC Press, a Taylor and Francis Group, Print ISBN: 9781926895314, eBook: 978-1-46-656887-7, **2013**, www.AppleAcademicPress.com

56. Hariri Asli, K.; Heat Transfer and Phase Transitions in Binary Mixtures of Liquid With Vapor Bubble, Water Hammer Research; Advances in Nonlinear Dynamics Modeling, PhD.; Thesis of Kaveh Hariri Asli, Toronto, Canada, Published by Apple Academic Press, Inc.; Exclusive worldwide distribution by CRC Press, a Taylor and Francis Group, Print ISBN: 9781926895314, eBook: 978-1-46-656887-7, **2013**, www.AppleAcademicPress.com

57. Hariri Asli, K.; Water Hammer and Surge Wave Modeling, Water Hammer Research; Advances in Nonlinear Dynamics Modeling, PhD.; Thesis of Kaveh Hariri Asli, Toronto, Canada, Published by Apple Academic Press, Inc.; Exclusive worldwide distribution by CRC Press, a Taylor and Francis Group, Print ISBN: 9781926895314, eBook: 978-1-46-656887-7, **2013**, www.AppleAcademicPress.com

58. Hariri Asli, K.; Computer Models for Fluid Interpenetration, Water Hammer Research; Advances in Nonlinear Dynamics Modeling, PhD.; Thesis of Kaveh Hariri Asli, Toronto, Canada, Published by Apple Academic Press, Inc.; Exclusive worldwide distribution by CRC Press, a Taylor and Francis Group, Print ISBN: 9781926895314, eBook: 978-1-46-656887-7, **2013**, www.AppleAcademicPress.com

59. Hariri Asli, K.; Computer Models for Heat Flow, Water Hammer Research; Advances in Nonlinear Dynamics Modeling, PhD.; Thesis of Kaveh Hariri Asli, Toronto, Canada, Published by Apple Academic Press, Inc.; Exclusive worldwide distribution by CRC Press, a Taylor and Francis Group, Print ISBN: 9781926895314, eBook: 978-1-46-656887-7, **2013**, www.AppleAcademicPress.com

60. Hariri Asli, K.; Heat Flow and Porous Materials, Water Hammer Research; Advances in Nonlinear Dynamics Modeling, PhD.; Thesis of Kaveh Hariri Asli, Toronto, Canada, Published by Apple Academic Press, Inc.; Exclusive worldwide distribution by CRC Press, a Taylor and Francis Group, Print ISBN: 9781926895314, eBook: 978-1-46-656887-7, **2013**, www.AppleAcademicPress.com

61. Hariri Asli, K.; Thermal Environment, Water Hammer Research; Advances in Nonlinear Dynamics Modeling, PhD.; Thesis of Kaveh Hariri Asli, Toronto, Canada, Published by Apple Academic Press, Inc.; Exclusive worldwide distribution by CRC Press, a Taylor and Francis Group, Print ISBN: 9781926895314, eBook: 978-1-46-656887-7, **2013**, www.AppleAcademicPress.com

62. Hariri Asli, K.; Heat Flow in Non-homogeneous Material, Water Hammer Research; Advances in Nonlinear Dynamics Modeling, PhD.; Thesis of Kaveh Hariri Asli, Toronto, C nada, Published by Apple Academic Press, Inc.; Exclusive worldwide distribution by CRC Press, a Taylor and Francis Group, Print ISBN: 9781926895314, eBook: 978-1-46-656887-7, **2013**, www.AppleAcademicPress.com

63. Hariri Asli, K.; Nagiyev, F. B.; Haghi, A. K.; Aliyev, S. A.; Physical and Numerical Modeling of Fluid Flow in Pipelines: A computational approach, International Journal of the Balkan Tribological Association, ISSN: 1310–4772, 16, *1,* Sofia, Bulgaria, 20–34, **2010**, Scientific Information Database (SID), 2013 http://www.sid.ir/en/ViewPaper.asp?ID=192583&varStr=2; ASLI%20KH.; NAGHIYEV%20F. B.; ALIYEV%20S. A.; HAGHI%20A. K.; JOURNAL%20OF%20THE%20BALKAN%20TRIBOLOGICAL%20ASSOCIATION%20; 2010; 16; 1; 20; 34

64. Moody, L. F.; "Friction Factors for Pipe Flow," Trans. ASME, **1944,** *66,* 671–684.

65. Bergeron L.; "Water hammer in Hydraulics and Wave Surge in Electricity," John Wiley and Sons, Inc.: NY, **1961,** 102–109.

66. Bracco A.; McWilliams, J. C.; Murante G.; Provenzale A.; Weiss, J. B.; "Revisiting freely decaying two-dimensional turbulence at millennial resolution,"*Phys. Fluids, 11,* **2000,** *12,* 2931–2941.

67. Loytsyanskiy, L. G.; Fluid, Moscow: Nauka, **1970,** 904.

68. Haghi, A. K.; Some Aspects of Microwave Drying, The Annals of Stefan cel Mare University, Year, **2000,** *2(14),* 22–25.

69. Haghi, A. K.; A Thermal Imaging Technique for Measuring Transient Temperature Field- An Experimental Approach, The Annals of Stefan cel Mare University, Year, **2000,** *6(12),* 73–76.

70. Haghi, A. K.; Experimental Investigations on Drying of Porous Media using Infrared Radiation, Acta Polytechnica, **2001,** *41(1),* 55–57.

71. Haghi, A. K.; A Mathematical Model of the Drying Process, Acta Polytechnica **2001,** *41(3),* 20–23.

72. Haghi, A. K.; Simultaneous Moisture and Heat Transfer in Porous System, Journal of Computational and Applied Mechanics 2001, *2(2),* 195–204.

73. Haghi, A. K.; A Detailed Study on Moisture Sorption of Hygroscopic Fiber, Journal of Theoretical and Applied Mechanics 2002, *32(2),* 47–62.

74. Flory, P. J.; Statistical Mechanics of Chain Molecules, Interscience Pub. NY, 1969.

75. Joukowski N.; Paper to Polytechnic Soc. Moscow, spring of1898, English translation by Miss O.; Simin. Proc. AWWA, **1904,** 57–58.

76. Wisniewski, K. P.; Design of pumping stations closed irrigation systems: Right. / Vishnevsky, K. P.; Podlasov, A. V. – Moscow: Agropromizdat, **1990,** 93.

77. Nigmatulin, R. I.; Khabeev, N. S.; Nagiyev, F. B.; Dynamics, heat and mass transfer of vapor-gas bubbles in a liquid. Int. J.; Heat Mass Transfer, vol.24, N6, Printed in Great Britain, **1981,** 1033–1044.

78. Vargaftik, N. B.; Handbook of thermo-physical properties of gases and liquids. Oxford: Pergamon Press, **1972,** 98.

79. Laman, B. F.; Hydro pumps and installation, **1988,** 278.

80. Nagiyev, F. B.; Kadyrov, B. A.; Heat transfer and the dynamics of vapor bubbles in a liquid binary solution. DAN Azerbaijani, S. S. R.; **1986,** *4,* 10–13.

81. Alyshev V. M.; Hydraulic calculations of open channels on your PC. – Part 1 Tutorial. – Moscow: MSUE, **2003,** 185.

82. Streeter V. L.; Wylie, E. B.; "Fluid Mechanics," McGraw-Hill Ltd.; USA, **1979,** 492–505.

83. Sharp B.; "Water hammer Problems and Solutions," Edward Arnold Ltd.; London, **1981,** 43–55.

84. Skousen P.; "Valve Handbook," McGraw Hill, New York, HAMMER Theory and Practice, **1998,** 687–721.

85. Shaking, N. I.; Water hammer to break the continuity of the flow in pressure conduits pumping stations: Dis. on Kharkov, **1988,** 225.

86. Tijsseling,"Alan E Vardy Time scales and FSI in unsteady liquid-filled pipe flow," **1993,** 5–12.

87. Wu, P. Y.; Little, W. A.; measurement of friction factor for flow of gases in very fine channels used for micro miniature, Joule Thompson refrigerators, Cryogenics **1983,** *24(8),* 273–277.

88. Song C. C. et al.; "Transient Mixed-Flow Models for Storm Sewers,"*J. Hyd. Div.;* Nov.; **1983,** *109,* 458–530.

89. Stephenson D.; "Pipe Flow Analysis," Elsevier, *19,* S. A.; **1984,** 670–788.

90. Chaudhry, M. H.; "Applied Hydraulic Transients," Van Nostrand Reinhold Co.; N. Y.; **1979,** 1322–1324.

91. Chaudhry, M. H.; Yevjevich, V. "Closed Conduit Flow," Water Resources Publication, USA, **1981,** 255–278.

92. Chaudhry, M. H.; Applied Hydraulic Transients, Van Nostrand Reinhold, New York, USA, **1987,** 165–167.

93. Kerr, S. L.; "Minimizing service interruptions due to transmission line failures: Discussion," Journal of the American Water Works Association, *41, 634,* July **1949,** 266–268.

94. Kerr, S. L.; "Water hammer control," Journal of the American Water Works Association, *43,* December **1951,** 985–999.

95. Apoloniusz Kodura, Katarzyna Weinerowska," Some Aspects of Physical and Numerical Modeling of Water Hammer in Pipelines," **2005,** 126–132.

96. Anuchina, N. N.; Volkov V. I.; Gordeychuk V. A.; Es'kov, N. S.; Ilyutina, O. S.; Kozyrev O. M. "Numerical simulations of Rayleigh-Taylor and Richtmyer-Meshkov instability using mah-3 code," J.; Comput. Appl. Math.**2004,** *168,* 11.

97. Fox, J. A.; "Hydraulic Analysis of Unsteady Flow in Pipe Network," Wiley, N. Y.; **1977,** 78–89.

98. Karassik, I. J.; "Pump Handbook – Third Edition," McGraw-Hill, **2001,** 19–22.

99. Fok, A.; "Design Charts for Air Chamber on Pump Pipelines,"*J. Hyd. Div.;* ASCE, Sept.; **1978,** 15–74.

100. Fok, A.; Ashamalla A.; Aldworth G.; "Considerations in Optimizing Air Chamber for Pumping Plants," Symposium on Fluid Transients and Acoustics in the Power Industry, San Francisco, USA, Dec, **1978,** 112–114.

101. Fok, A.; "Design Charts for Surge Tanks on Pump Discharge Lines," BHRA 3rd Int. Conference on Pressure Surges, Bedford, England, Mar.; **1980,** p.23–34.

102. Fok, A.; "Water hammer and Its Protection in Pumping Systems," Hydro technical Conference, CSCE, Edmonton, May, **1982,** 45–55.

103. Fok, A.; "A contribution to the Analysis of Energy Losses in Transient Pipe Flow," PhD.; Thesis, University of Ottawa, **1987**, 176–182.
104. Hariri Asli, K.; Nagiyev, F. B.; Water Hammer and fluid condition, Ministry of Energy, Gilan Water and Wastewater Co.; Research Week Exhibition, Tehran, Iran, December, **2007**, 132–148, http://isrc.nww.co.ir.
105. Hariri Asli, K.; Nagiyev, F. B.; Water Hammer analysis and formulation, Ministry of Energy, Gilan Water and Wastewater Co.; Research Week Exhibition, Tehran, Iran, December, **2007**, 111–131, http://isrc.nww.co.ir.
106. Hariri Asli, K.; Nagiyev, F. B.; Water Hammer and hydrodynamics instabilities, Interpenetration of two fluids at parallel between plates and turbulent moving in pipe, Ministry of Energy, Guilan Water and Wastewater Co.; Research Week Exhibition, Tehran, Iran, December, **2007**, 90–110, http://isrc.nww.co.ir.
107. Hariri Asli, K.; Nagiyev, F. B.; Water Hammer and pump pulsation, Ministry of Energy, Guilan Water and Wastewater Co.; Research Week Exhibition, Tehran, Iran, December, 2007, 51–72, http://isrc.nww.co.ir.

INDEX

Printed in the United States
by Baker & Taylor Publisher Services